Conservation Biology
with RAMAS® EcoLab

Susanne M. Shultz
The State University of New York at Stony Brook

Amy E. Dunham
The State University of New York at Stony Brook

Karen V. Root
Applied Biomathematics

Sheryl L. Soucy
The State University of New York at Stony Brook

Steven D. Carroll
The State University of New York at Stony Brook

Lev R. Ginzburg
The State University of New York at Stony Brook

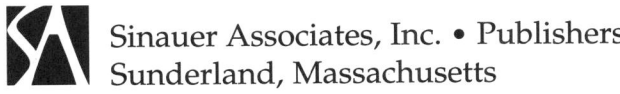

Sinauer Associates, Inc. • Publishers
Sunderland, Massachusetts

The Cover
Yukon, Canada. Boreal forest ecosystem
Photograph by Amy Dunham, Applied Biomathematics. © 1999 Applied Biomathematics
Design by Wendy Beck

Conservation Biology with RAMAS® EcoLab

© 1999 by Sinauer Associates, Inc.
All rights reserved.
This book may not be reproduced in whole or in part for any purpose whatever without permission from the publisher.

For information or to order, address:
    Sinauer Associates, Inc.
    P.O. Box 407
    Sunderland, Massachusetts 01375-0407 U.S.A.
    Fax: 413-549-1118, Internet: publish@sinauer.com

RAMAS® EcoLab Software
    © 1990–1995 Electric Power Research Institute
    © 1990–1999 Applied Biomathematics
    RAMAS is a registered trademark and Applied Biomathematics is a registered servicemark of Applied Biomathematics. Other companies and trademarks mentioned herein are trademarks of their respective companies.
    Amy Dunham, Susanne Shultz, and Karen Root created the original artwork throughout the text. © 1999 by Applied Biomathematics.

ISBN 0-87893-768-4

Manufactured in the United States of America

5 4 3 2 1

Acknowledgments

This laboratory manual was the collaborative effort of a number of people. The authors would especially like to acknowledge the creative and technical assistance of H. Resit Akçakaya and William Root, authors of RAMAS® EcoLab software at Applied Biomathematics, and Mark Burgman, an Associate Professor and Reader at the University of Melbourne, Australia.

Table of Contents

Preface vii

Laboratory 1 Biodiversity: The Diversity of Life 1
  Introduction 2
  What Is Biodiversity? 3
  Prelab Questions 6
  Sampling Biodiversity 8
  EXERCISE A: Quadrat sampling 9
  EXERCISE B: Invertebrate biodiversity 10
  EXERCISE C: Was your sample size sufficient? 14
  References 17

Laboratory 2 From Bacteria to Blue Whales: Growth Without Limits 18
  Introduction 19
  Growth in an Unlimited Environment 19
  Factors That Affect Population Growth 20
  Exponential Growth of a Natural Population 21
  EXERCISE A: Relating growth rate to change in population size 23
  EXERCISE B: Modeling exponential growth 23
  EXERCISE C: The influence of mortality factors 25
  EXERCISE D: Modeling blue whale recovery 27
  References 30

Laboratory 3 Competition in Osprey, Fish, and Barnacles: Limits to Population Growth 31
  Introduction 32
  Growth in a Limited Environment 32
  Types of Density Dependence 33
  EXERCISE A: Scramble competition among fish 36
  EXERCISE B: Contest competition in osprey 38
  EXERCISE C: Ceiling model of density dependence for barnacles 41
  References 43

Laboratory 4 Wood Storks and Honeyeaters: Estimating Population Characteristics 44
  Introduction 45
  Estimating Population Abundance 45
  Age and Stage Classes 46
  Survival Rate 47
  Measuring Fecundity 48
  What Is a Life Table? 49
  What Is a Matrix? 49
  EXERCISE A: Counting wood storks 50
  EXERCISE B: The helmeted honeyeater 54
  References 60

Laboratory 5   Grizzly Bears: The Problems of Small Populations   61
Introduction   62
Environmental and Demographic Variation   62
Loss of Genetic Variation in a Population   63
Extinction Vortexes and Allee Effects   64
Population Viability Analysis   64
Grizzly bear populations   65
EXERCISE A. Viability of several grizzly bear populations   66
EXERCISE B: Running multiple simulations   70
References   71

Laboratory 6   Giant Pandas: Risks Faced by Endangered Species   72
Introduction   73
Indicators, Flagships, and Keystone Species   73
Endangered Species   74
Life History and Population Size   75
Threats to Endangered Species   76
Giant Pandas in China   78
EXERCISE A. Growth in a panda population   79
EXERCISE B. The impact of poaching on pandas   82
References   84

Laboratory 7   Hector's Dolphins and the Red-Cockaded Woodpecker: Conserving Dwindling Populations   85
Introduction   86
Where Should We Focus Conservation Efforts?   86
Saving the Loggerhead Sea Turtle   86
EXERCISE A: A wildlife sanctuary for Hector's dolphins   88
EXERCISE B: Conservation of the red-cockaded woodpecker   92
References   96

Laboratory 8   African Market Hunting and Tuna Exploitation: Maintaining Sustainable Levels of Harvesting   97
Introduction   98
The Goal of Sustainable Harvesting   98
Harvesting Practices   99
Poaching: A Threat to Sustainable Harvesting   101
Calculating Maximum Sustainable Yield   101
EXERCISE A: Harvesting wild game in Africa   102
EXERCISE B: Harvesting bluefin tuna   105
References   113

Laboratory 9   The African White Rhino: Too Many for Their Own Good?   114
Introduction   115
Limiting Population Growth   115
Elephants of Zimbabwe   117
EXERCISE A: White rhinos at Umfolozi Reserve   118
EXERCISE B: Management of an overpopulated reserve   121
EXERCISE C: Balancing a declining population   123
Reference   126

Laboratory 10 The Wild Ass and the Black Footed Ferret: Reintroduction of Endangered Species 127
    Introduction 128
    Time and Uncertainty 128
    Predictions and Error 128
    Environmental Stochasticity 128
    Releasing Individuals of Endangered Species 130
    EXERCISE A: The Asiatic wild ass 132
    EXERCISE B: The black-footed ferret 139
    References 145

Laboratory 11 Park Size and Species Diversity: Lessons from Islands 146
    Introduction 147
    Area and the Number of Species 148
    Island Biogeography and Reserve Design 149
    Fragmentation and Edge Effects 151
    Applications in Conservation Planning 152
    EXERCISE A: The species-area relationship 152
    EXERCISE B: Predicting species abundance 155
    References 158

Laboratory 12 Rescuing the Spotted Owl: Conserving Species in Multiple Populations 159
    Introduction 160
    How Do Metapopulations Form? 160
    Correlation among Populations 161
    EXERCISE A: The southern California spotted owl 162
    EXERCISE B: Designing reserves for the spotted owl 167
    References 171

Laboratory 13 Biodiversity's Biggest Threat: Human Population Growth 172
    Introduction 173
    How Many People? 174
    A Human-Dominated Planet 174
    Overconsumption 175
    Declining Growth Rates? 177
    EXERCISE A: Human population growth 178
    EXERCISE B: The effectiveness of immigration control programs 181
    EXERCISE C: Fertility control programs 183
    References 185

Laboratory 14 The Case of Patrick's Marsh Wren: Making Decisions to Protect Species 186
    Introduction 187
    Conservation Planning 187
    Patrick's Marsh Wren 188
    The Scenario 188
    EXERCISE Evaluating options for Patrick's marsh wren 190
    Worksheets 204

Glossary 209

Using RAMAS EcoLab 211
    Installation 211
    Uninstalling RAMAS EcoLab 211
    Running the Programs 211

Preface

"The diversity of life forms, so numerous that we have yet to identify most of them, is the greatest wonder of this planet."
E. O. Wilson, editor's forward in *Biodiversity*, 1988,
National Academy Press, Washington, D.C.

The diversity of life, or biodiversity, of this planet is now at risk of being destroyed. Habitat loss or modification is regarded as the major threat for most of the world's threatened and endangered species. As humans continue to extract resources, build structures, and alter the landscape, the conflict between human needs and those of native (nonhuman) species will escalate. There is a great need for trained scientists and managers that can evaluate the potential impacts and propose policy and management solutions for these problems. Conservation biology is emerging as an interdisciplinary approach that provides the theoretical and empirical tools to study biodiversity and its threats.

Conservation biology, a relatively new field, is rapidly evolving to meet the needs of the human community as it struggles to resolve the ongoing conflict with the natural world. New techniques such as population viability analysis have been developed to quantify the risks of extinction that populations may face. This laboratory manual explores many of these concepts and methods that are being used to today to actively manage endangered and threatened species, including population viability analysis and island biogeography.

This set of fourteen labs is designed as a teaching tool for undergraduate environmental or ecological laboratory courses or possibly for advanced high school courses in biology. Real problems in ecology and conservation biology are explored, using RAMAS EcoLab software, that can be understood by students even without strong mathematical backgrounds. Major concepts are explored through examples based on real data from threatened and endangered species. Although these exercises are not designed to be a comprehensive analysis for any population, they demonstrate the techniques and applications possible using models. Students are guided through the process of conservation from collection of the data, parameter estimation, and population viability analysis to the development and evaluation of management alternatives. The series of exercises culminates in class participation in a court case using all of the skills learned in the previous labs.

In this manual we explore a number of important topics that are relevant to environmental studies:
- Biodiversity
- Population growth
- Population parameters (demography)
- Density dependence
- Extinction
- Conservation of threatened and endangered species
- Sustainable harvesting
- Management of wildlife
- Reintroduction of endangered species
- Island biogeography
- Metapopulation biology

❖ Human population growth and consumption
❖ Conservation decision-making

Our technical emphasis in these labs is the use of models, mathematical representations of the natural world, to address issues related to biodiversity and the conservation of species. Models are a particularly useful tool because they not only allow the user to test out different scenarios but also to examine underlying assumptions and their effects on the outcome.

Throughout the lab manual students use the Windows-based, user-friendly program RAMAS EcoLab to build and evaluate models of various types (i.e., single population, stage-based, and metapopulation models). Use of RAMAS EcoLab is designed to make modeling approachable. Students are not required to learn the complex mathematical techniques such as matrix algebra and can concentrate, instead, on learning how data are used and how demographic parameters such as survival and fecundity are estimated. This approach also allows the students to explore the effects of individual parameters and uncertainty in general.

Each lab provides background material, emphasizing the important concepts, a series of exercises using data from actual studies of endangered and threatened species, and questions for the students to answer. The manual also provides clear step-by-step instructions on the use of RAMAS EcoLab and the provided sample files. There is a separate section "Using RAMAS EcoLab" that explains the functions of the program and how to operate it. Not only does the lab manual explore the latest concepts in conservation, but also it also utilizes the same technology that professionals in the field use to tackle these issues.

We have designed these labs to complement most textbooks in conservation biology. This volume can either be used independently as the basis of a laboratory course, or as a complement to a lecture course based on any of the texts as shown in the following table, which highlights the relevant chapters as they relate to the laboratory exercises in this book.

Laboratory	Meffe & Carroll, Second Edition	Primack, Second Edition	Akçakaya, Burgman, Ginzburg	Bolen & Robinson	Caughley & Gunn	Caughley & Sinclair	Hunter
1	4,5	1,2,4,5,6		2,4,22	1,3,4	1,12	3,7
2	7		1	5	13	4	
3			1,3	7,8,9		5,6,8	
4		11	2,4,5		5,6	12,14	
5		11,20,21,22	2,7	22	12	15	16
6	6,15,16,17	8,11	3,7		5,6	14,15	10
7	6	10,12	5,7	2,3,10	8,11	14,15,16	9,11
8	11,12,13	10,20,21	3,8	2,10,20	11	16	9,14
9	15,18	15,16,17	8	16,21	10	14,15	13
10	11,12,13,14	13,14,19	7,8	16,21	8,13	15	11,12
11	9,10	9	6,7	16	9,10	14,15	8,13
12	7		6	5		5	
13		9,10	4				14,15
14	15,16,17	15	6,7,8	20,22	11,12	15	14,15,16

Laboratory 1
Biodiversity: The Diversity of Life

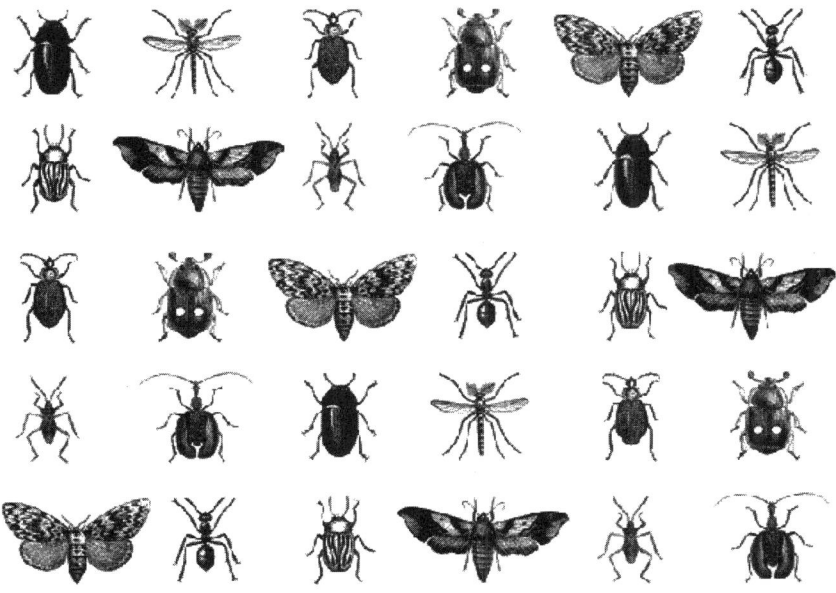

Accelerating rates of loss of biodiversity due to human impact have been documented worldwide. Many people are concerned that a loss of the diversity of ecosystems and species will affect the enjoyment, health, and standard of living of future generations. In this laboratory you will learn about biodiversity, its importance, and how it is measured.

Introduction

Biological diversity, or biodiversity, refers to the variety of living things and the ecosystems that they form. Issues relating to the conservation of biodiversity have been attracting a lot of attention recently from conservation biologists, policy makers, and the general public. Biodiversity increases when new species evolve (speciation) and decreases when species become extinct. Every species alive today may play an important role in the earth's ecosystem. The loss of a single plant species can cause the extinction of many animals and insects that depend upon it.

Humans benefit directly from the diversity of species because it provides food, medicine, building products, and clothing. Even obscure species, with which we have no direct contact, can indirectly affect our everyday lives. This is because of the intricate web of interactions among all species on Earth. Since the beginning of civilization, people have depended on natural resources for their livelihood and health. Species that are yet undiscovered by scientists may some day become important new resources. It was recently discovered that the rosy periwinkle, a plant that inhabits the forests of Madagascar, can provide anti-cancer drugs for childhood leukemia and Hodgkin's disease.

Human activity has caused a loss of biodiversity on all levels, and some biologists believe that the earth may be on the verge of a mass extinction. Habitat destruction, overharvesting, and introduction of exotic (introduced) species have caused the demise of thousands of species worldwide.

Ecologists have estimated that between 3 and 20 million species inhabit the Earth today. There are more species alive today than there have been at any other time in Earth's history. So far, only about 1.75 million of these species have been identified, and many will become extinct before they receive a name. The number of species disappearing through extinction is also greater than ever before. The present rate of extinction may be as high as 20,000 species per year. That means that by the year 2050 one-third of all the world's species could be lost.

Biodiversity has been increasing since life began, approximately 3.2 billion years ago. However, the fossil record has shown that the increase in biodiversity has been punctuated by 5 major extinction events (Figure 1.1). If extinction is a natural process, and biodiversity has recovered from it in the past, why should we be concerned about extinctions occurring today? The loss of biodiversity resulting from the activity of a single species (humans) is unprecedented and may not be reversible. In the past, the evolutionary processes that create new species eventually exceeded the loss of species. Today's extinction rate greatly exceeds the rate at which species are replaced by speciation and if present activity of humans continues it seems unlikely that this trend will reverse.

If the loss of biodiversity continues at its present rate, it will not only affect proper functioning of ecosystems, but it will be harmful to humans as well. The loss of species may be due to direct actions such as hunting or to indirect actions such as habitat loss. As human populations and their consumption of natural resources continue to grow, the rate at which biodiversity is being lost will increase unless action is taken to conserve the remaining species or curb human growth and consumption. Figure 1.2 illustrates the increase in the number of mammal and bird extinctions as the human population size has increased.

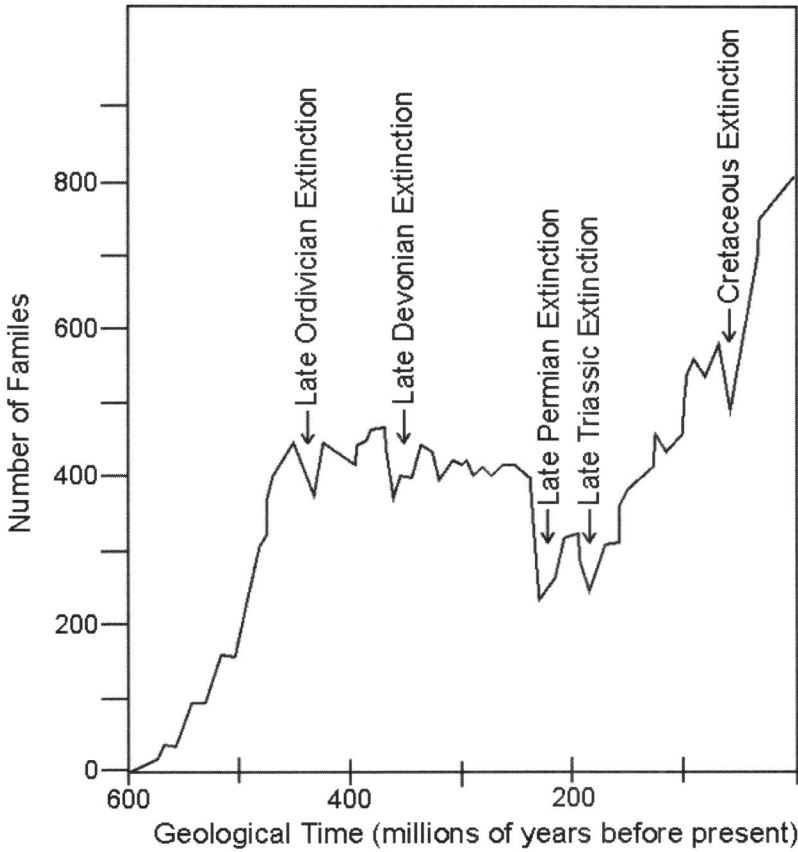

Figure 1.1. Graph depicting the change in the number of marine animal families of species over time and the 5 major extinction events in Earth's history (after Sepkoski 1984).

What Is Biodiversity?

The term "biodiversity" literally means the variability of life. Depending on how the term is used, it can take on several different meanings. Biodiversity may refer to the number of community types within a geographical region, in which case we use the term "habitat biodiversity." It may refer to the amount of genetic variation within a species, which is "genetic biodiversity," or it may refer to the number of species in a given area, termed "species biodiversity."

Habitat biodiversity refers to the variety of ecosystems in an area. It includes all of the different communities that compose the ecosystem, the physical patterning of communities, and their functions such as nutrient cycling and energy flow. Communities of organisms (plants, animals, fungi, insects, etc.) function together as units, called ecosystems. Communities can be very large such as all of the organisms in Lake Superior, medium sized such as the organisms that inhabit a patch of oak forest, or small such as the plants and animals in a little pond.

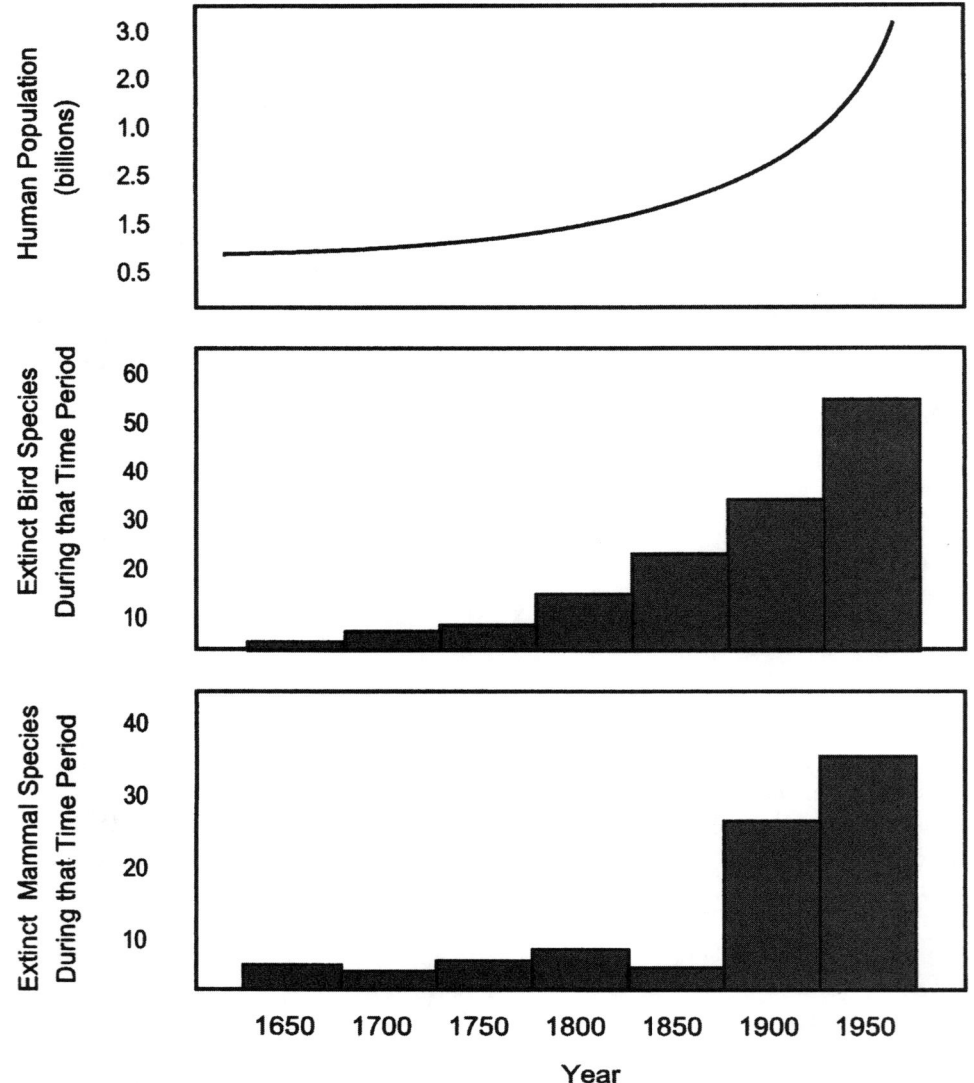

Figure 1.2. As human populations increase so do the number of bird and mammal species that become extinct (after Cunningham and Saigo 1995).

Genetic biodiversity accounts for much of the difference we see between individuals of the same species. It refers to the variation in the genetic makeup (DNA) of individuals within a population, subspecies, or species. The different breeds of dogs that are all members of the same species represent a good example of genetic diversity within a species. The genetic makeup of an individual greatly affects its physical and biochemical characteristics and how it will react to the environment. Therefore, having some genetic variation in a species or population means that not all individuals will be equally susceptible to problems that might be associated with a changing environment or a new disease. Individuals in small populations are often prone to inbreeding (mating with related individuals) because there are very few individuals from which to select a mate. Inbreeding results in a loss of genetic diversity within that population. For many species, close inbreeding also results in offspring that are unhealthy because of developmental or other problems.

When people talk of biodiversity, they are most commonly referring to species biodiversity, the variety of species in an area. Taxonomists define a species in two ways. The first is the biological species concept in which a species consists of all individuals that potentially can mate and produce fertile offspring. However, there are occasional problems with using this concept, such as when a taxonomist wants to identify a new specimen or an organism that reproduces asexually (without mating). An alternative definition is the morphological species concept. By this definition, a species is a group of individuals that is morphologically, physiologically, or biochemically distinct from other groups. The difficulty of this definition is that all individuals within a species differ somewhat, and all species seem to have different amounts of intra-specific (within a species) variation. Therefore, it is difficult to judge how different a group of organisms must be before it should be considered a separate species. The biological species concept is easy to define, so it is typically used when distinguishing one well-known species from another. However, for asexually reproducing species, or species that are not well known, the morphological species concept must be used.

Species biodiversity includes two distinct components. The first is species richness or, simply, the total number of species in a region. The second is species evenness, which refers to the degree to which the abundance of each of the species is similar. For example, imagine two communities, each containing a total of 100 individuals that belong to two species. In one community there are 50 individuals of species A and 50 individuals of species B. In the second community there are 2 individuals of species A and 98 individuals of species B. If biodiversity meant only species richness, both communities are equally diverse because they both have two species. If, however, you define biological diversity to include both species evenness and richness, the two communities would have different biodiversity values. The second community would be considered less diverse than the first community because rare species lower the biodiversity estimate of a community when species evenness is considered.

Ecologists often use mathematical expressions to express ideas in numerical form, called indices (singular: index). A commonly used index of biodiversity that includes both species evenness and richness is the Simpson index.

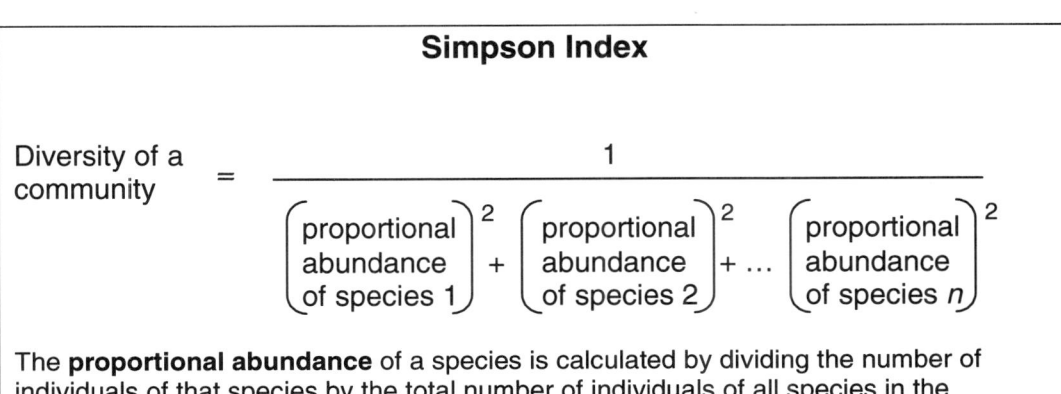

Simpson Index

$$\text{Diversity of a community} = \frac{1}{\left(\begin{array}{c}\text{proportional}\\\text{abundance}\\\text{of species 1}\end{array}\right)^2 + \left(\begin{array}{c}\text{proportional}\\\text{abundance}\\\text{of species 2}\end{array}\right)^2 + \ldots \left(\begin{array}{c}\text{proportional}\\\text{abundance}\\\text{of species } n\end{array}\right)^2}$$

The **proportional abundance** of a species is calculated by dividing the number of individuals of that species by the total number of individuals of all species in the community.

There has been much controversy concerning the use of biodiversity indices such as the Simpson index. The concept of biodiversity is very complex, and one cannot expect a single index to capture all the information about an ecological community. Many important biological aspects are lost in any single index, such as which species are present and which are the most abundant. Still, diversity indices can be descriptive and useful if one does not expect them to encompass all of the important aspects that species biodiversity represents. People have come up with many different indices that all differ in how much they emphasize species richness and evenness. Choosing the right one often depends on your goal and what aspects of biodiversity are important for answering a particular question.

Prelab Questions

1. Describe a situation in which it might be appropriate for a biologist to use only species richness as a measure of biodiversity.

2. Describe a situation in which it might be appropriate for a biologist to use a biodiversity index that incorporates both species richness and species evenness.

3. Examine Figure 1.3. Which community has the greater species richness?

Community 1

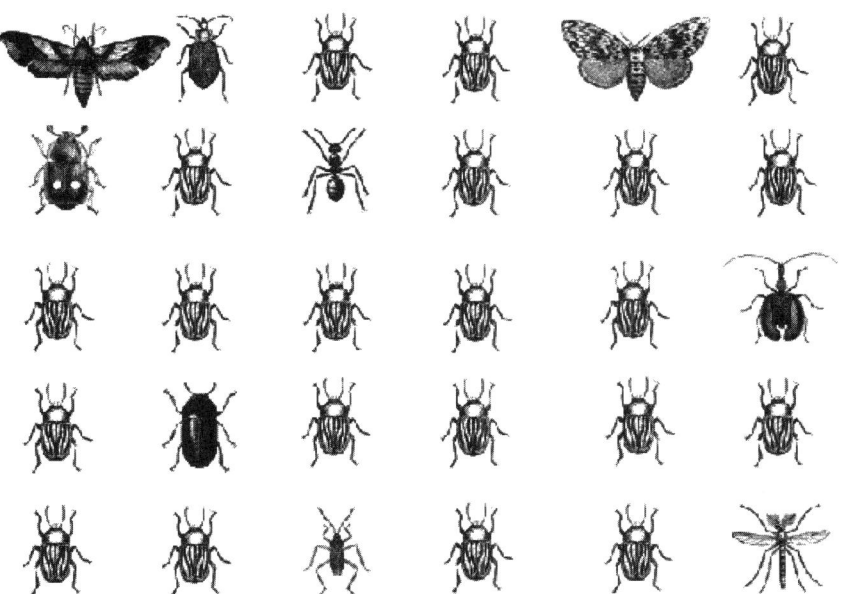

Community 2

Figure 1.3. Two insect communities illustrate the difference between species evenness and richness.

4. Use the Simpson index to determine which community has the greater diversity when both species richness and evenness are included.

Sampling Biodiversity

It is virtually impossible to count every individual or even every species in a region. Rare species, in particular, may be very difficult to find. For these reasons biodiversity values do not rely on absolute counts (counting every individual or every species in an area). Instead biologists estimate the number of individuals and species by sampling the environment. Unfortunately, estimates can be easily biased by the amount of effort an ecologist puts into the census of an area. The more sampling one does, the more individuals one will likely find. If a biologist wishes to compare biodiversity estimates for a number of communities, it is important that he or she use the same sampling method in each community.

One way to determine if a sampling effort is adequate is to construct a species accumulation curve. To do this, one plots the total number of species found against the number of samples taken. The total number of species found should level off with greater sampling effort. When sampling begins, new species are encountered quickly. As more and more samples are taken, many of the same species are reencountered and fewer new species are left that can be found. An estimate of species richness is attained when the curve levels off, or becomes horizontal. The optimum number of samples is one that reasonably measures the total number of species, without wasting effort by oversampling. In Figure 1.4, 8 to 10 samples are adequate for estimating the total number of species in this hypothetical community. Twenty samples would require twice as much time and would produce approximately the same estimate.

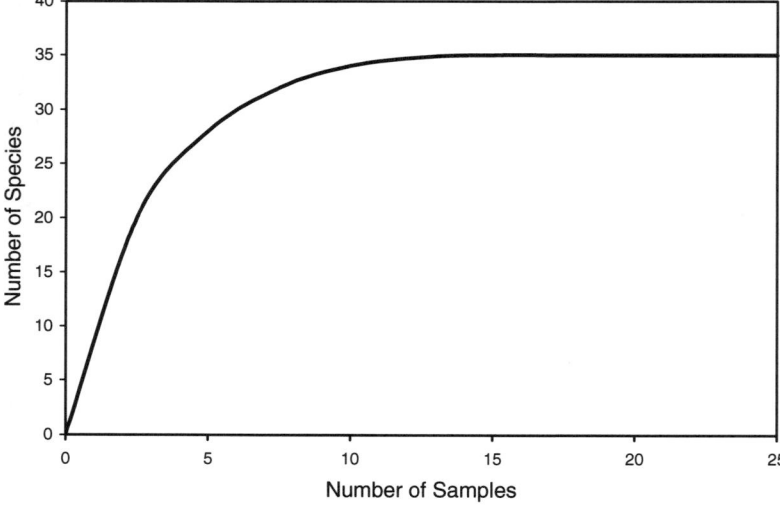

Figure 1.4. Hypothetical species accumulation curve.

Measuring and monitoring biodiversity are important for making appropriate management and policy decisions. These methods are needed for the proper documentation of newly discovered species and monitoring the status of ecosystems and individual species over time. Creating inventories and monitoring biodiversity are also needed to determine how certain human activities affect biological diversity and how we can minimize their effects.

In this lab you will have the opportunity to compare the species biodiversity of the soil invertebrate community in two different habitats. Invertebrates are animals that do not have an internal skeleton. The invertebrates found in the soil include insects, spiders, millipedes, centipedes, various types of worms, and some mollusks. The invertebrate communities that inhabit the soil may not be the most conspicuous but they are some of the most diverse in the world. You will learn methods to sample soil invertebrates and to compare the species richness and evenness of two habitats.

Your instructor has selected two areas that are ecologically different, such as a forest and a field. These areas will be your two study sites for comparing the biological diversity of invertebrates in two habitat types. If you will be collecting soil samples, begin with Exercise A. You should expect to complete all of Exercise A and Exercise B through step 4 in one lab period. You will finish Exercise B in the next lab period. Begin with Exercise B if the soil samples have already been collected by your instructor or if sampling the soil community in you area is not an option.

Exercise A: Quadrat sampling

List of required materials for Exercise A

Each group will need:
- Pen and paper to record observations
- A metric ruler
- 4 10 cm (3.9 inches) lengths of string
- A small hand shovel
- Two plastic bags
- Labels to be placed on the bags

Exercises

1. Once you reach a study site, describe the habitat that you see around you. Write down some basic observations about the physical characteristics of the habitat that might affect the soil community. For example, is the soil moist or dry? Can sunlight reach the ground? Is the area open or covered with vegetation? What is the soil type (e.g., sand, clay, organic material)? Also record some biological characteristics of the area that might affect what lives in the soil. What kinds of plants are within your study site? Is there leaf litter covering the soil (you might want to measure how thick the litter layer is)? Do you think there are any predators that might affect the invertebrate community such as birds or insectivorous mammals that live in the area?

2. Pick an area on the ground and mark the corners of a square measuring 10 × 10 cm (3.9 inches × 3.9 inches) with the 4 lengths of string you've brought. This area that you have marked off is called a quadrat.
3. Remove the top 10 cm of soil from the quadrat. You can use the ruler to check how deep you are digging. You and your classmates should be very careful to remove the same amount of soil from each quadrat so that you can compare the results.
4. Place the soil into a plastic bag that you have labeled with the site name, date, and the initials of everyone in your group.
5. Repeat this procedure in the other habitat type.

Exercise B: Invertebrate biodiversity

Background

It is possible that sampling the soil community in your area is not a viable option. If this is the case, we recommend sampling the "biodiversity" of a surrogate community, such as a jar of jelly beans or a mixture of dried beans. Two different mixtures can represent the two different ecosystems found at "site 1" and "site 2." Groups should sample these communities by using a standard-size collecting tool such as a measuring cup. Each color jelly bean or type of dried bean will represent a different "species." Follow the instructions below, beginning with step 7.

List of required materials for Exercise B

Each group will need:
- A plastic soda bottle with the bottom cut off
- Hardware cloth or plastic mesh (approximately 4 holes per inch)
- A jar, labeled with the group number
- Heavy tape
- A solution of 70% ethanol with a drop of dishwashing soap
- A small lamp (unless sunlight is available)
- A pair of tweezers
- A magnifying glass or dissecting scope, if insects are small
- An insect field guide (if available, not absolutely necessary)

Exercises

1. To separate the invertebrates from the soil, you will create what is called a Tullgren funnel. Cut the bottom out of a plastic soda pop bottle and then place a piece of cloth mesh or wire mesh inside the bottle (see Figure 1.5).

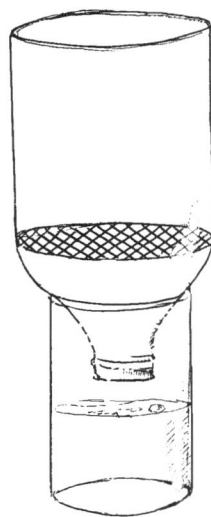

Figure 1.5. A Tullgren funnel constructed from a plastic bottle for processing soil samples.

2. Label a jar with your group's initials and fill it one-third full with a solution of 70% ethanol and a drop of dishwashing soap. The soap prevents the small insects from simply walking on top of the water and escaping. Instead, they sink and are quickly killed by the ethanol solution.
3. Place the jar underneath the Tullgren funnel. Secure the funnel with tape and place one soil sample on the mesh.
4. Direct a lamp over the sample or place it in direct sunlight. As the soil warms and begins to dry out, the invertebrates will crawl down deeper into the soil in search of moisture. Eventually they will reach the bottom of the soil and fall through the mouth of the bottle into the ethanol solution. Allow a minimum of 1½ hrs for this step.
5. Next separate the invertebrates by pouring the ethanol solution through a fine sieve.
6. Carefully remove the invertebrates from the sieve using tweezers and sort them into categories based on their morphology. You may need a magnifying glass or dissecting scope for this step. Use insect field guides if they are available, or simply use general classifications (black ants, red ants, pill bugs, different beetle types, earthworms, centipedes, etc.). You may want to create a reference collection to ensure that everyone in the class uses the same classification scheme.
7. Count the individuals in each category.
8. You do not want to bias the order in which samples are recorded on the blackboard; therefore, your instructor will randomly assign each group a number before the you write your results on the board. Data should be recorded in the order of the sample number assigned (group 1 will write its data on the board first). Each group of students should write the species category and species abundance values on the board under the headings "site 1" and "site 2." Make sure to specify which group collected the data. In the next few steps you will combine data from all groups.
9. Summarize the class data and fill out Table 1.1. Use a separate sheet of paper if you need more categories than provided.

Table 1.1 Abundance of various species groups at sites 1 and 2.

Species Group	Abundance in Site 1	Abundance in Site 2

10. Create two bar graphs with the class data, one for each site. Figure 1.6 is an example of a bar graph. The vertical axis (y-axis) of the graph represents the abundance, or number of items, in each category. The horizontal axis (x-axis) is used for the categories you have measured (in your case, species groups). To begin, label the x-axis "Species Group" and the y-axis "Abundance." Plot the data from your table.

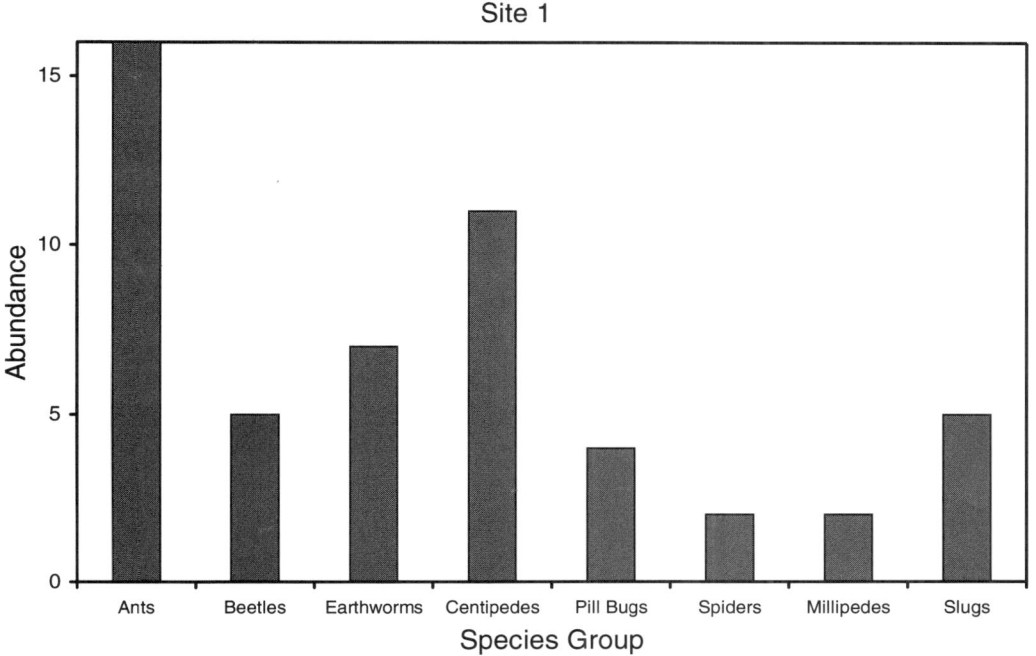

Figure 1.6. A graph comparing the number of individuals of 8 species groups at one site.

11. Determine which site has the higher species richness by comparing the total number of species groups found in each site.

12. Calculate the Simpson index for your group's data from site 1 and from site 2.

13. Calculate the Simpson index for the total class data from site 1 and site 2.

Exercise C: Was your sample size sufficient?

Background

Now you will determine whether the sample size your class used was adequate for estimating the species group biodiversity in the two sites. To do this you will construct a species accumulation curve. Remember, each group took a sample at each site, so there should be one sample per site for each group. For example, if you broke the class into 10 groups, there should be 10 samples for each of the two sites, for a total of 20 samples.

Exercises

1. To help you construct a species accumulation curve for site 1, fill out Table 1.2. Use a separate sheet if necessary for additional samples. For each sample, you will write the number of *new* species categories found in that sample (i.e., species that were not observed in any of the previous samples). For sample 1 you will write down the number of species categories that were found in sample 1. For sample 2 you write down the number of species categories found in sample 2 that were *not* observed in sample 1 (for example, you would write the number 3 if pill bugs, earthworms, and black ants were found in sample 2 but not in sample 1). For sample 3 you will record the number of species categories that were found in sample 3 but not found in *either* sample 1 or 2 (for example, you would write the number 2 if millipedes and slugs were found in sample 3 but not in samples 1 or 2). Continue this process for all samples, writing down only the number of new species categories found for each sample.

Table 1.2. Summary of new species groups and cumulative number of species groups for site 1.

Sample Number	Number of Species Found That Were Not in Any Previous Samples	Cumulative Number of Species Groups
1		
2		
3		
4		
5		
6		
7		
8		
9		
10		

2. In the next column you will calculate the *cumulative* number of species found with successive samples. For each sample you will write down the *total* number of species found up through that sample. For example, let's say you found 8 species groups in sample 1 and 3 new species groups in sample 2 (i.e., 3 species groups that were not present in sample 1). The cumulative number of species for sample 1 is 8, and for sample 2 it is 8 + 3 = 11. The two new species categories found in sample 3 would be added to that number (8 + 3 + 2 = 13). Continue this process for samples 1 through 10.
3. Repeat steps 1 and 2 for the data from site 2, filling in Table 1.3.

Table 1.3. Summary of new species groups and cumulative number of species groups for site 2.

Number of Samples	Number of Species Found That Were Not in Any Previous Samples	Cumulative Number of Species Groups
1		
2		
3		
4		
5		
6		
7		
8		
9		
10		

4. You now will use the data from Tables 1.2 and 1.3 to create species accumulation curves (refer to Figure 1.4). Label your *x*-axis "Number of Samples" and your *y*-axis "Cumulative Number of Species Groups." Plot the data from the last column of Table 1.2 for site 1 on your graph. Connect the points you have drawn. Use the same procedure to create another curve on the same graph for site 2. Make sure you clearly label each line.
5. Based on your species accumulation curve, do you feel confident estimating the total number of species groups at each site? Decide as a group if your class has taken a sufficient number of samples for estimating the invertebrate biodiversity for these two sites. Be prepared to discuss your answer.

Questions

1. What are the major types of biological diversity? Describe them briefly.

2. What is the difference between species richness and species evenness?

3. How did the estimate of invertebrate biodiversity (the Simpson index) differ when you used your group's data and when you used the class data as a whole?

4. Why is it better to take many samples when estimating biodiversity?

5. In which site did you observe the highest invertebrate species richness? Which site has the greatest species diversity when you consider richness and evenness?

6. What physical and biological factors do you think contributed to the differences in species composition and biodiversity that you found between sites 1 and 2?

7. List some of the reasons it might be important to conserve biodiversity.

Your lab report should include the following:

1. Answers to Prelab questions
2. Two bar graphs of species abundances (sites 1 and 2)
3. Biodiversity assessment of invertebrates in the two sites (steps 11, 12, and 13 in Exercise B)
4. Species accumulation curve including sites 1 and 2
5. Answers to questions 1 through 7

References

Cunningham, W. P., and B. W. Saigo. 1995. *Environmental Science: A Global Concern.* Third Edition. Brown, Dubuque, IA.

Primack, R. B. 1998. *Essentials of Conservation Biology.* Second Edition. Sinauer Associates, Sunderland, MA.

Sepkoski, J. J., Jr. 1984. A kinetic model of Phanerozoic taxonomic diversity III. Post-Paleozoic families and mass extinctions. *Paleobiology*, 10: 246–267.

Wilson, E. O. 1988. *Biodiversity.* National Academy Press, Washington, D.C.

Laboratory 2
From Bacteria to Blue Whales: Growth Without Limits

Understanding population growth is an important aspect of the study of living systems. If we are to conserve species and biodiversity, we must have a good understanding of how and why populations change over time. In this laboratory you will learn some fundamental concepts of population ecology through the use of computer models.

Introduction

Every living organism is a member of a **population** and species may be composed of one, several, or many populations. Just as an organism is made up of individual cells, a population is made up of individual organisms. All of the organisms of a single species living in one area constitute a population. We can speak, for example, about the spotted owl population in the Pacific Northwest or of the bacteria population in a rotting apple. Populations, like organisms, can grow in size, move over a geographic range, consume resources, and interact with other populations. Just as an entire organism has properties that individual cells do not, a population has unique characteristics and functions not shared by solitary organisms. Because there is a distinction between a population and the organisms within it, it is important for us to gain an understanding of how populations, not just organisms, behave.

We can use information about how a population changes over time to estimate whether the population will increase, decrease, or remain stable. There are many practical applications of this field in the areas of fisheries, wildlife management, forest harvest planning, and the protection and management of threatened species. For example, it may be important for a state agency to estimate the number of deer available for hunting. Population modeling is a tool that predicts the future abundance of a population based on its past birth and death rates. Using this tool, a state wildlife officer could estimate the number of deer likely to survive that year and calculate how many hunting permits should be issued.

Growth in an Unlimited Environment

All populations can change in size depending on how many individuals are added or removed. A population can increase in size through either birth or **immigration**. A population can decrease through either death or **emigration**. If births and immigration are adding new individuals at the same rate that death and emigration are removing them, a population will remain at a constant size. Most often, however, these rates are not equal within one time period, so populations undergo periods of increase and decrease.

If food or other resources do not limit a population of organisms, the natural pattern of growth will follow a simple progression called **exponential growth**. During exponential growth the number of individuals increases at a constant percentage at each time step. When abundance is plotted against time, the result is a J-shaped growth curve as seen in Figure 2.1.

Perhaps the easiest way to visualize this growth process is to imagine how population growth may occur from a single bacterial cell (Figure 2.2). In the first generation one bacterial cell will reproduce by splitting into two cells. In the next generation each of the two cells splits to create a total of 4 bacterial cells. Those 4 will become 8 and then sixteen, thirty-two, and so on. After each generation the population has increased by 100%, producing an exponential growth curve like Figure 2.1.

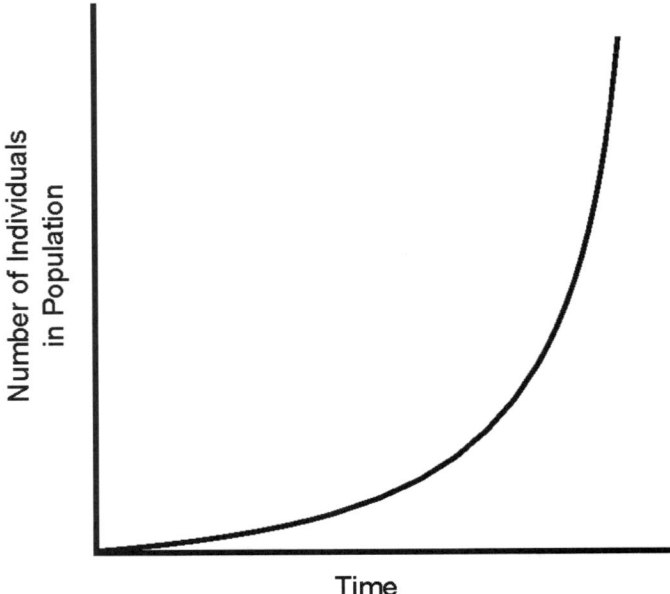

Figure 2.1. Exponential growth in an unlimited environment.

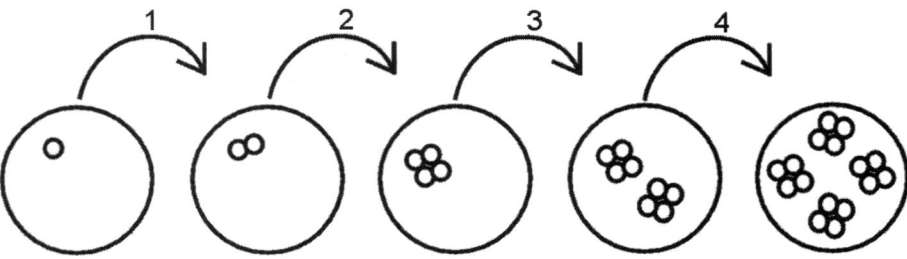

Figure 2.2. Growth of bacteria in 4 generations where the population doubles at each time step.

Factors That Affect Population Growth

In the bacteria example above, population growth is affected only by the birth rate of the population. However, populations change in size not only by the addition of organisms through birth and immigration but also by the removal of individuals through death and emigration. When death and emigration occur, a population does not necessarily lose its potential to grow exponentially. If the birth and immigration events combine to exceed the death and emigration events, the population may still grow in an exponential fashion.

> Change in population size =
> (births + immigration) − (deaths + emigration)

The change in population size is related to another measure, the growth rate. The growth rate is defined below:

> **Population size at the next time step =
> (Growth rate) × (Population size at the present step)**

You can see from this equation that if the growth rate is greater than 1, the population will grow. Similarly, if the growth rate is less than 1, the population will decline. For example, if the current population size (of some imaginary population) is 100 and the growth rate is 1.10, the population will grow to 1.10 × 100 = 110 by the next time step. Note that the change in population size in this case is +10, which is a positive number. However, if the growth rate is 0.90, the population will decline to 0.90 × 100 = 90 by the next time step. Here, the change in population size is −10, which is a negative number. If the growth rate is equal to 1, the population size will not change.

In summary, a population's growth rate is a combination of all the factors causing the population to grow and to decline. If we speak of the growth rate of a particular population, we are not implying that the population is growing. A population has a growth rate greater than 1 only when births and immigration are greater than death and emigration, in which case the population will increase. A population has a growth rate less than 1 when births and immigration are lower than deaths and emigration, in which case the population will decline. When the number of births and immigrants equals the number of deaths and emigrants, the population maintains a constant number of total individuals. Keep this in mind throughout the lessons that follow.

Exponential Growth of a Natural Population

A good example of exponential growth occurring in nature is the muskox population in Alaska, following its reintroduction. The muskox is a large woolly mammal (Figure 2.3) whose natural range once included the colder regions of North America and Greenland. During the 1700s and 1800s, excessive hunting caused a population decline in North America. The last North American muskox was killed in Alaska in the 1850s.

In 1936, the Alaskan legislature decided to obtain animals from a Greenland population and reintroduce them onto Alaska's Nunivak Island. Nunivak Island was considered ideal habitat for the muskox because it had plenty of food resources and was free of predators such as wolves or grizzly bears. If the population became established, biologists planned to use individuals from the island to subsequently reintroduce the species to the mainland. Because the area had been free of muskoxen for a long time, food was plentiful, and the growing population experienced little competition. As a result of the favorable environmental conditions, the birth rate (number of offspring born per individual) was high and the mortality rate, or death rate, was low. The population increased exponentially in the first 40 years (Figure 2.4).

Figure 2.3. The muskox was reintroduced onto Alaska's Nunivak Island almost ninety years after its disappearance.

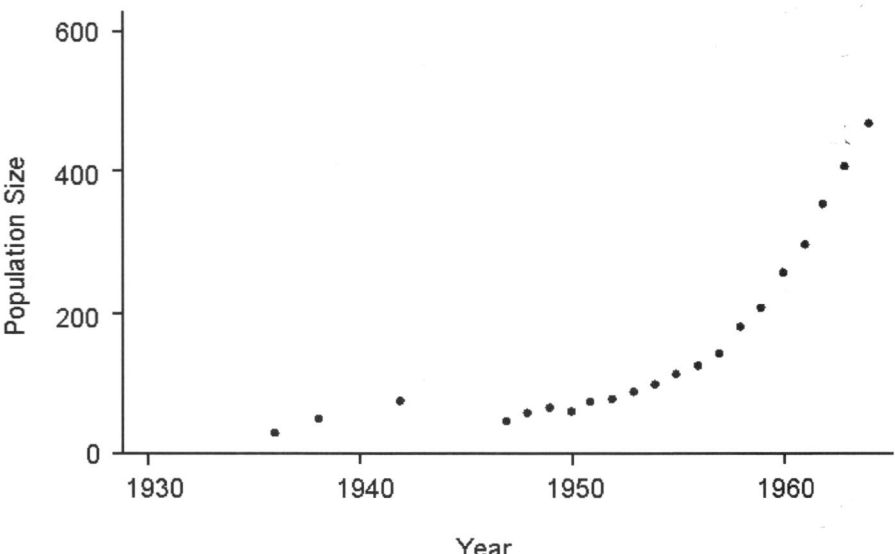

Figure 2.4. Population growth of muskoxen on Nunivak Island based on data from Spencer and Lensink 1970, adapted from Akçakaya et al. 1999.

Exercise A: Relating growth rate to change in population size

1. If the population size of widgets (some imaginary organism) in some area is 1000 and the growth rate is 1.26, what will the population size be at the next time step? What is the change in population size (remember to specify whether the change is positive or negative)?

    1000 × 1.26 = 1260

    260

    positive

2. If the growth rate is 0.67, what will the population size be at the next time step? What is the change in population size?

    (1000)(.67) = 670

    −330

3. If the population size of widgets is 1000 at one time step and grows to 1230 by the next time step, what is the growth rate? What is the change in population size?

    1000x = 1230

    1.23

4. If the population growth rate is 1.00, what will be the change in population size by the next time step? 0

Exercise B: Modeling exponential growth

Background

In the next exercises you will explore the dynamics of exponential population growth using RAMAS EcoLab. First, consider a bacterial population that begins as just a single individual. This individual undergoes asexual reproduction to produce two organisms after a period of 20 minutes. These two individuals can, after an additional 20 minutes, each reproduce again to result in a total of 4 bacteria. This process continues so that after each reproduction there are not just *two* more individuals, but *double* the number that were alive in the preceding time step.

Exercises

1. Open the program RAMAS EcoLab (double-click on its icon with your mouse).
2. Click on *Population Growth* (single population models).
3. Under the <u>M</u>odel menu select *General <u>I</u>nformation*. In that window, title this file *Exponential growth of bacteria*.
4. Also in the *General <u>I</u>nformation* window, change the <u>D</u>uration from 0 to 10. By doing this, you have set the program to simulate ten time steps of reproduction. In this exercise, each time step is 20 min. long.
5. Also in the *General <u>I</u>nformation* window, set the number of *Replications* to 1. You will run one simulation to view how the size of the bacteria population is expected to change over time. Click OK to exit this window.
6. Under the <u>M</u>odel menu select *<u>P</u>opulation*. Change the <u>I</u>nitial Abundance to 1. The bacteria population begins with a single individual.
7. Also in the *<u>P</u>opulation* window, enter a *Survival Rate* of 0.0 and a *Growth Rate* of 2.00. By doing this you are indicating that after one time step the original individual no longer exists and is instead replaced by two new individuals. You do not need to enter a value for birth rate because the program will calculate birth rate as the growth rate minus the survival rate (this will be discussed in a later exercise). Click OK to exit this window.
8. Now you are going to run the simulation; under the <u>S</u>imulation menu select <u>R</u>un, or press *Ctrl-R* on your keyboard. At the bottom right corner of this window you will see a message when the simulation is complete. At this point you may close the window by clicking on the X in the upper right corner of the window.
9. Under the *Results* menu, select *<u>T</u>rajectory Summary*. Here you can view the population growth curve. The time scale is in 20 min. time steps. Print the plot of the *Trajectory Summary* by clicking on the printer icon.
10. You can view the numbers corresponding to the plot by clicking the *Show Numbers* button in the upper left corner of the *<u>T</u>rajectory Summary* window. In this table of numbers, you will see 5 columns, labeled as follows:

    Minimum   −1 S.D.   Average   +1 S.D   Maximum

    These columns have different meanings only when you run more than one replication. When you do this, you can see the smallest population size at a given time step for all the replications (minimum), the largest population size at a given time step (maximum), and the average population size plus or minus one standard deviation (−1 S.D. and +1 S.D.). Standard deviation is a unit to measure how much variability there is among the different replications. However, because you ran only one replication here, each column will have the same number within each time step.
11. You can return to the plot by clicking the *Show Plot* button, also in the upper left corner of the *<u>T</u>rajectory Summary* window.
12. Answer the questions that follow. Exit the *<u>T</u>rajectory Summary* by clicking on the X in the upper right corner of the window.

Questions

1. How many individuals did this single organism produce after ten time steps?

    1024

2. Recalling that each time step is 20 minutes long, how many minutes (or hours) did it take to generate 1000 bacteria?

    $$\frac{200}{1024} = \frac{x}{1000} \quad = 1024x = 200,000$$

Exercise C: The influence of mortality factors

Background

In the preceding example you considered only birth as a factor that affected population growth, and you ignored death and migration. Now you will add the effects of mortality. You are still ignoring migration in this exercise.

Suppose there is a population of 5000 chipmunks living in a forest. Under optimal conditions, these small mammals will reproduce rapidly, so in 1 month's time, another 4 individuals are added for every 10 individuals alive. The birth rate can be computed as 4/10 = 0.40 offspring per individual per month. However, during that same time period there is one death out of every 10 members of the population. Therefore, the individual survival rate is 9/10 = 0.90. The survival rate is equal to 1.0 minus the mortality rate, which is 0.10 in this example. The population growth rate is defined as follows (assuming no migration):

R	=	f	+	s
population growth rate		fecundity (birth rate)		survival

Recall that you multiply the growth rate by the current population size to get the next population size.

Exercise

1. Open a new file in the *Population Growth* (single population models) window of RAMAS EcoLab, by clicking on the *New* file icon in the upper left corner of the window. You will be asked if you want to save the file from the previous exercise, but you need not do so.

2. The *General Information* window from the *Model* menu will automatically open. In that window, title this file *Growth of a chipmunk population*.
3. Also in the *General Information* window, change the *Duration* from 0 to 10. By doing this, you have set the program to simulate ten time steps. In this exercise, each time step is 1 year long.
4. Also in the *General Information* window, set the number of *Replications* to 1. We will run one simulation to view how the size of the chipmunk population is expected to change over time. Click OK to exit this window.
5. Under the *Model* menu select *Population*. Change the *Initial Abundance* to 5000. The chipmunk population in the forest is estimated to have 5000 individuals.
6. Also in the *Population* window, enter a *Survival Rate* of 0.90 and a *Growth Rate* of 1.30. Recall that the birth rate (f) is 0.40 (40%) and the survival rate (s) is 0.90 (90%), so the growth rate (R) can be calculated by $R = f + s = 0.40 + 0.90 = 1.30$. Click OK to exit this window.
7. Now you are going to run the simulation; under the *Simulation* menu select *Run*, or press *Ctrl-R* on your keyboard. At the bottom right corner of this window you will see a message when the simulation is complete. At this point you may close the window by clicking on the X in the upper right corner of the window.
8. Under the *Results* menu, select *Trajectory Summary*. Here you can view the population growth curve. The time scale is in 1 year time steps. Print the plot of the *Trajectory Summary* by clicking on the printer icon.
9. You can view the numbers corresponding to the plot by clicking the *Show Numbers* button in the upper left corner of the *Trajectory Summary* window. You can return to the plot by clicking the *Show Plot* button, also in the upper left corner of the *Trajectory Summary* window.
10. Answer the questions that follow. Exit the *Trajectory Summary* by clicking on the X in the upper right corner of the window.

Questions

1. What was the birth rate (f) for the bacteria simulation and what was the birth rate (f) for the chipmunk population?

2. What was the mortality rate (= 1.0 minus survival) for the bacteria population and what was the mortality rate for the chipmunk population?

3. Do you observe exponential growth in the chipmunk simulation?

4. What is the growth rate (R) for the bacteria population? For the chipmunk population?

5. How does the pattern of growth for the chipmunks compare with that for the bacteria (is it the same, faster, or slower)?

6. Under what conditions should exponential growth occur?

Exercise D: Modeling blue whale recovery

Background

The blue whale is perhaps the largest animal ever to have lived, measuring up to 33m (110 ft) and weighing up to 17,000 kg (38,000 lb) (Figure 2.5). There are 3 distinct populations of blue whales, which live in the North Atlantic, North Pacific, and the southern hemisphere. The two northern populations do not come into contact with each other because the oceans are separated by land. The northern and southern populations are isolated from one another because each migrates to the tropics during the cold months of their respective hemisphere. Prior to human exploitation, 90% of the blue whales lived in the southern hemisphere, which held between 150,000 and 210,000 blue whales. Most of the whaling pressures in the past two centuries have focused on this population.

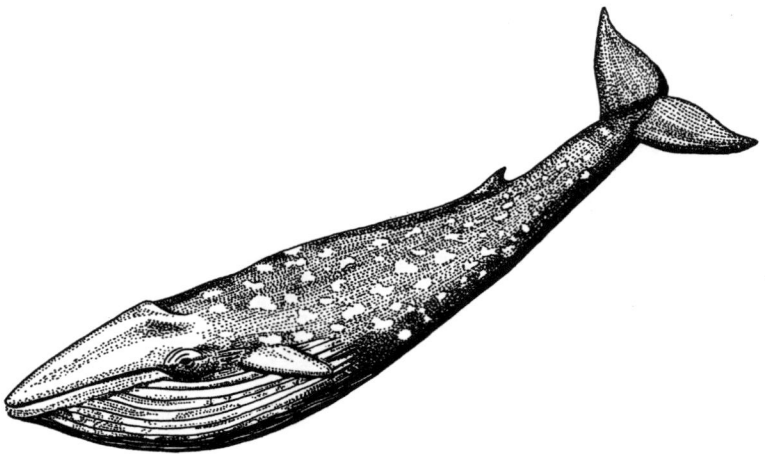

Figure 2.5. The blue whale.

In the mid-1920s whaling technology was developed that gave the whaling ships much greater efficiency. So began the decline of the blue whale. The whales were hunted for their meat and for blubber to make oil for fuel and lamps. After little over 10 years of hunting, the population had dropped to 20,000 to 50,000 individuals. By 1965 only 14,000 individuals remained, and many countries agreed to stop hunting for fear of the blue whale's extinction.

Despite the decreased hunting pressures, the blue whale has had trouble recovering from its small population size. The refusal of some countries to stop hunting and other pressures such as the increasing harvest of krill, which are small shrimp (the whale's principal food source), have hindered the blue whale's recovery. It was estimated in 1989 that only about 500 blue whales remained in the oceans of the southern hemisphere. Recent and more optimistic reports, however, have suggested that the population size is probably increasing.

If krill fishing is controlled and human hunting pressures are relieved, the habitat may be more favorable for recovery. It has been estimated that the blue whale would have a survival rate of 0.96 per year under these conditions. That means each whale would have a 96% chance of surviving to the next year. With an increase in survival as well as fecundity, the whale population is expected to have a growth rate between 1.02 and 1.10 per year. These figures mean that the population abundance would increase between 2 and 10% each year. In this exercise you will simulate the recovery of the blue whale population using the software program RAMAS EcoLab.

Exercises

1. Open a new file in the *Population Growth* (single population models) window of RAMAS EcoLab by clicking on the *New* file icon in the upper left corner of the window. You will be asked if you want to save the file from the previous exercise, but you need not do so.

2. The *General Information* window from the *Model* menu will automatically open. In that window, title this file *The southern blue whale population*.
3. Also in the *General Information* window, change the *Duration* from 0 to 20. By doing this, you have set the program to simulate 20 time steps of reproduction. In this exercise, each time step is 1 year long.
4. Also in the *General Information* window, set the number of *Replications* to 1. We will run one simulation to view how the size of the blue whale population is expected to change over time. Click *OK* to exit this window.
5. Under the *Model* menu select *Population*. Change the *Initial Abundance* to 500. This is the estimated size of the blue whale population in the southern oceans in 1989.
6. Also in the *Population* window, enter the *Survival Rate* given above that corresponds to the estimated survival rate with natural mortality.
7. Also in the *Population* window, enter the *Growth Rate* given above that corresponds to the lowest predicted growth rate if whales are relieved from pressures of hunting and depleted food sources. Click *OK* to exit this window.
8. Now you are going to run the simulation; under the *Simulation* menu select *Run*, or press *Ctrl-R* on your keyboard. At the bottom right corner of this window you will see a message when the simulation is complete. At this point you may close the window by clicking on the X in the upper right corner of the window.
9. Under the *Results* menu, select *Trajectory Summary*. Here you can view the population growth curve. The time scale is in 1 year time steps. Print the plot of the *Trajectory Summary* by clicking on the printer icon.
10. You can view the numbers corresponding to the plot by clicking the *Show Numbers* button in the upper left corner of the *Trajectory Summary* window. You can return to the plot by clicking the *Show Plot* button also in the upper left corner of the *Trajectory Summary* window. Click *OK* to exit this window.
11. Now you are going to run the model using a different value for the growth rate. Under the *Model* menu select *Population*. Change the growth rate to the highest predicted growth rate value for whales relieved from pressures of hunting and depleted food sources. Run the simulation and print your results by repeating steps 8 through 10. Click *OK* to exit this window.

Questions

1. How many years will it take for the population to double (reach 1000 individuals) with the highest growth rate?

2. How many years will it take for the population to double (reach 1000 individuals) with the lowest growth rate?

3. How does the growth rate affect your predictions about the impact of harvesting? What implications does that have for protection of the blue whales?

4. Discuss one reason you think modeling or predicting population growth might be important for conserving a species such as the blue whale.

Your lab report should include the following:

1. Answers to questions 1 through 4 for Exercise A
2. A trajectory summary graph and answers for questions 1 and 2 for Exercise B
3. A trajectory summary graph and answers for questions 1 through 6 for Exercise C
4. Two population trajectory graphs and answers to questions 1 through 4 for Exercise D

References

Akçakaya, H. R., M. A. Burgman, and L. Ginzburg. 1999. *Applied Population Ecology using RAMAS®EcoLab.* Second Edition. Sinauer Associates, Sunderland, MA.

Meffe, G. K. and C. R. Carroll. 1997. *Principles of Conservation Biology.* Second Edition. Sinauer Associates, Sunderland, MA.

Spencer, D. L. and C. J. Lensink. 1970. The muskox of Nunivak Island, Alaska. *Journal of Wildlife Management,* 34:1–15.

Laboratory 3
Competition in Osprey, Fish, and Barnacles: Limits to Population Growth

Natural populations usually do not grow exponentially because various features of their environment limit them. Limitations, whether they are competition over food, nesting sites, or other factors, influence the dynamics of a population. In this laboratory you will learn about density dependence: what the causes are, how populations respond to it, and what effect it has on the shape of a population growth curve.

Introduction

All populations have the capacity for exponential growth in an unlimited environment. However, because environments are not unlimited, populations cannot continue to increase indefinitely (Begon et al. 1996). As a population increases, resources are used up faster than they are replaced. Individuals within a population must compete with each other for these limited resources. The habitat that a species occupies may become so overcrowded or degraded that it can no longer support a population undergoing exponential growth. Not only does crowding cause competition for resources, but an overcrowded population may pollute the environment with its own wastes or attract predators or disease.

When various species are faced with overcrowding, their reactions may differ, but all reactions result in a lowering of the growth rate (R). Recall that the growth rate of a population depends upon birth and immigration as well as death and emigration. A reduction in fecundity and immigration will lower the growth rate. A lower growth rate can also result from higher mortality or emigration. When death and emigration exactly compensate birth and immigration, the average growth rate of a population is 1. At this point the population size remains fairly constant. The population generally stabilizes at a density that the environment is able to support (i.e., with enough food and shelter and not too much competition and predation). This density is known as the **carrying capacity** of the population (K).

Growth in a Limited Environment

Because the factors leading to a growth rate of 1 depend upon the changing density of individuals, they are called **density dependent** factors (Akçakaya et al. 1999). The negative effects on population growth will become more pronounced as the density of individuals in a population increases. Density-dependent factors limit the growth of a population in several ways.

Dispersal

If dispersal (emigration) is possible, surplus individuals may leave to look for a less populated area. For many animals, the dispersal rate increases as the density of individuals increases. Not only will more individuals disperse with increased crowding, but a higher percentage of the total population will disperse.

Plants generally are not mobile, but individuals may disperse as seeds by wind, water, or even by animals. Frugivores, or fruit eaters, may be strongly attracted to areas with high densities of fruiting plants. The frugivores help to disperse seeds by eating the fruit and then depositing the seeds with their feces after they leave the area. Although this does not thin the existing plants, proportionately fewer seedlings will begin to grow around the parent plants, and the seeds may be deposited where it is less crowded.

Increased mortality

Sometimes dispersal is not possible because a population is isolated by a barrier such as a mountain or river, or a less crowded area is not available. Individuals that are not free to leave, such as plants or other nonmobile organisms, may be faced with a crowded environment. When this occurs, resources become scarce and wastes accumulate, causing the environment to become less favorable. Individuals competing for scarce resources such as food, water, sunlight, or nutrients may die. Because of the accumulated wastes and the frequent contact between individuals in a crowded environment, parasites and diseases often spread quickly.

In the same manner that an area with a lot of fruit will attract many frugivores, areas with high densities of prey will attract large numbers of predators. A population of predators may also grow when prey species are at high densities because more food is available. More predators will in turn cause the survival rate of the prey population to decrease. For all of these reasons, the mortality rate of a population often increases as the density of individuals increases.

Decreased reproduction

The number of offspring produced for many species depends on the available resources. As resources become scarce, individuals may reproduce less or even cease to reproduce altogether. For example, the number of red squirrels born in a litter depends on the condition of the mother. If food is plentiful and the mother is able to store a lot of fat in her body, she will produce, on average, more young than during times when resources are scarce. Nonfood resources such as nesting or denning sites may also be important for reproduction. A female may not be able to raise her young if other individuals are using all available sites. As densities increase, resources necessary for reproduction become less available, and the average reproductive rate of each individual is likely to decline.

Types of Density Dependence

How a population responds to density dependent factors depends on the ecology of the species and on the resources that become limiting. For the rest of this discussion we will assume that individuals cannot migrate into or out of the population.

Scramble competition

As the size of a population increases due to the births of new individuals, resources are used up. If these scarce resources are shared more or less equally among individuals, there will not be enough resources for any one individual. Density dependent factors will affect all individuals equally. This type of competition is called **scramble competition**.

Scramble competition usually occurs when an organism's limiting resource is dispersed evenly across a habitat, rather than in patches that are easy to defend. This leads to leads to equal sharing of resources because it is difficult for an individual or group of individuals to monopolize the resource when it becomes scarce. Consider competition for food among newly hatched fish in a pond. If there are very few young, all of them will have enough food to survive. However, if the density of young is above the carrying capacity, all of the individuals will be smaller or weaker due to insufficient resources. At very high densities, conditions for the whole population may be so poor that no individuals survive to reproduce. Grazers such as the rhinoceros provide another example of scramble competition. It is difficult for individuals to defend a field of grasses, so all rhinos are affected by overgrazing when their densities get very high.

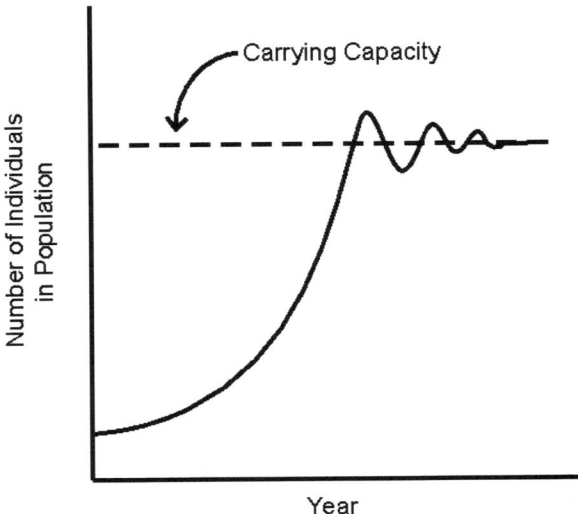

Figure 3.1. An example of the effect of scramble competition on population growth.

The effects of scramble competition on population growth over time depend on how quickly the population would normally grow without density dependence (i.e., the maximum growth rate). In populations with a low maximum growth rate, the effect of density dependence is felt before the population reaches its carrying capacity. In such a case, the growth rate of the population gradually declines toward 1 (no growth) as it approaches its carrying capacity. With a very high maximum growth rate, a population may exceed its carrying capacity before density dependence takes effect. This may result in fluctuations or oscillations in the population abundance that will differ in magnitude and shape depending on the strength of the density dependence and the maximum growth rate of the population (Figure 3.1).

Contest competition

When individuals in a population do not share resources equally, they exhibit **contest competition**. Even in the most crowded conditions, there are some individuals who get enough resources to survive and reproduce by defending those resources. Competition for territories may be a form of contest competition. A territory is an area that an individual, or group of individuals, defends to control the access to resources found within the territory boundaries. The owner(s) of a territory will have a good chance of both surviving and reproducing. Individuals that are unable to obtain a territory will either disperse to another area, remain but fail to breed, or in some cases even die.

When a population is at low density, survival and reproductive rates are high, so the population grows quickly. However, when conditions become crowded, some individuals are unable to obtain enough resources. This decreases the average fecundity and survival of the population. The effects of contest competition are apparent before a population reaches its maximum density. The growth rate of the population gradually declines toward 1 as the carrying capacity is reached (Figure 3.2). Theoretically, there will never be negative growth (population decline) because a certain number of individuals will always able to secure adequate resources.

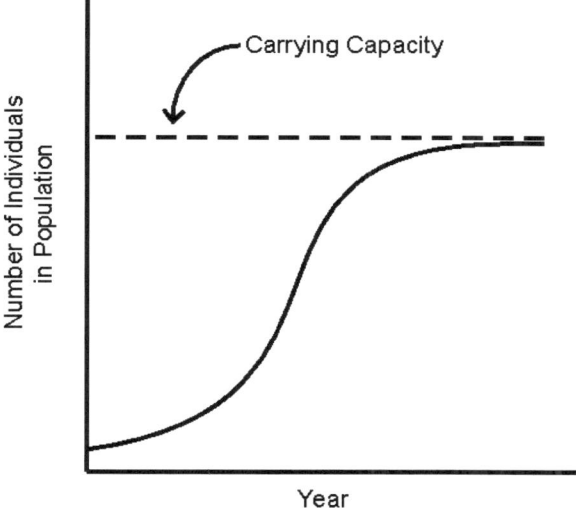

Figure 3.2. Effects of contest competition on population growth.

Ceiling model of competition

A third type of density dependence is called the ceiling model. Here the population will increase exponentially until it reaches its carrying capacity. At this point, growth will stop and the population will remain at that level (Figure 3.3). In this case, the carrying capacity acts as a population ceiling over which the population cannot increase. A population ceiling may result from an extreme case of contest competition. This may occur where individuals fight for a territory and losers either disperse or die. Once the resources are claimed, the population can no longer grow.

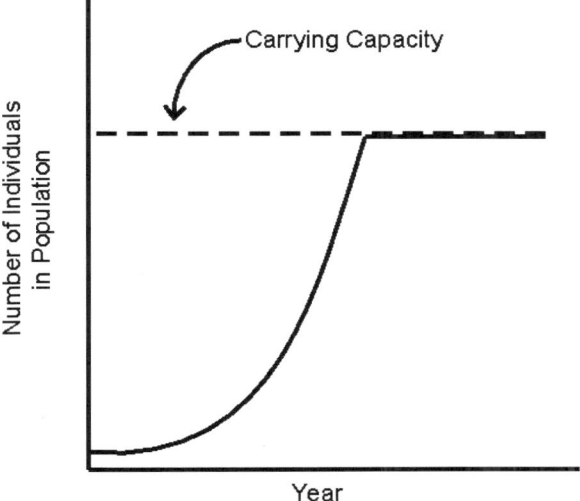

Figure 3.3. Ceiling type of density-dependent growth.

In the following exercises you will examine the three different types of density dependence by simulating the growth patterns of three different populations using RAMAS EcoLab.

Exercise A: Scramble competition among fish

Background

In this exercise, you will simulate scramble competition within a population of sunfish in a lake in New Hampshire. The fish feed on plankton, and as the population increases, all individuals are affected. Using scramble competition and different initial abundances, you will evaluate the effects of density dependence on a small population of fish.

Exercises

1. Open the program RAMAS EcoLab (double-click on its icon with your mouse).
2. Click on *Population Growth* (single population models).
3. Under the *Model* menu select *General Information*. In that window, title this file *Scramble competition in sunfish*.
4. Also in the *General Information* window, change the *Duration* from 0 to 20. By doing this, you have set the program to simulate 20 time steps of reproduction. In this exercise, each time step is 1 year long.
5. Under the *Model* menu select *Population*. Change the *Initial Abundance* to 10. The hypothetical sunfish population begins with 10 individuals in the lake. This abundance is well below the carrying capacity of the lake.
6. Also in the *Population* window, enter a *Growth Rate* of 2.50. This is the growth rate that the population will experience without the effects of density dependence.
7. To investigate the effect of scramble competition, select *Scramble* in the pull-down menu titled *Density Dependence Type*. Notice that the default type of density dependence is *Exponential*. This represents no density dependence, so exponential growth can occur indefinitely. Also in the *Population* window, set the carrying capacity to 500. Click *OK* to exit this window.
8. Now you are going to run the simulation; under the *Simulation* menu select *Run*, or press *Ctrl-R* on your keyboard. At the bottom right corner of this window you will see a message when the simulation is complete. At this point you may close the window by clicking on the X in the upper right corner of the window.
9. Under the *Results* menu, select *Trajectory Summary*. Here you can view the population growth curve. The time scale is in one-year time steps. Print the plot of the *Trajectory Summary* by clicking on the printer icon.
10. You can view the numbers corresponding to the plot by clicking the *Show Numbers* button in the upper left corner of the *Trajectory Summary* window. You can return to the plot by clicking the *Show Plot* button also in the upper left corner of the *Trajectory Summary* window.
11. In Table 3.1, record the abundance of the population at time steps 0, 5, 10, 15, and 20. Your answers should be recorded in the column corresponding to this simulation, which has an initial abundance of 500. Exit the *Trajectory Summary* by clicking on the X in the upper right corner of the window.
12. Now you are going to run the model using an initial abundance near the carrying capacity of the lake. Under the *Model* menu select *Population*. Change the initial

abundance to 450. Run the simulation and print your results by repeating steps 8 through 11. As in the previous step, record the abundance of the population at time steps 0, 5, 10, 15, and 20 in the appropriate column of the table.

13. Repeat the procedure for an initial abundance of 700 individuals to see what happens when you start with more individuals than the lake can support.

Table 3.1. Abundance of fish over time using the same carrying capacity but different values of initial abundance (N_i).

	Abundance		
Time	$N_i = 10$	$N_i = 450$	$N_i = 700$
0			
5			
10			
15			
20			

Questions

1. With an initial abundance of 10, how long does it take the population to reach 500 individuals?

2. With an initial abundance of 450, how long does it take the population to reach 500 individuals?

3. What happened to the fish population over time when the initial abundance was above the carrying capacity of the lake?

4. How do you think fish may be altering their environment when they are living at very high densities?

Exercise B: Contest competition in osprey

Background

In this exercise, you will use RAMAS EcoLab to simulate the growth of an osprey population. Osprey are raptors that breed on saltwater wetlands on the coast of North America and Europe. They are very vulnerable to human development because the dead trees on which they nest are often removed. About 20 years ago there were only about 30 osprey left on all of Long Island, New York. A conservation effort was made to save the populations by placing artificial platforms on the wetland areas. Only the females in this species that secure a nest are able to reproduce. Although the other birds cannot reproduce, they can still find food and will survive. Assuming that the environment is constant, their growth after the artificial platforms are put in place should resemble contest competition.

Exercises

1. Open a new file in the *Population Growth* (single population models) window of RAMAS EcoLab, by clicking on the *New* file icon in the upper left corner of the window. You will be asked if you want to save the file from the previous exercise, but you need not do so.
2. The *General Information* window from the *Model* menu will automatically open. In that window, title this file *Contest competition in osprey*.
3. Also in the *General Information* window, change the *Duration* from 0 to 20. By doing this, you have set the program to simulate 20 time steps of reproduction. In this exercise, each time step is 1 year long.
4. Under the *Model* menu select *Population*. Change the *Initial Abundance* to 30. Twenty years ago, the Long Island osprey population was estimated to have only thirty individuals.
5. Also in the *Population* window, enter a *Growth Rate* of 1.30. This is the growth rate that the population will experience until the effects of density dependence are felt.
6. To investigate the effects of contest competition, select *Contest* in the pull-down menu titled *Density Dependence Type*. Also in the *Population* window, set the carrying capacity to 300. Assume that artificial nesting sites have been set up which increase the carrying capacity to 300.
7. Now you are going to run the simulation; under the *Simulation* menu select *Run*, or press *Ctrl-R* on your keyboard. At the bottom right corner of this window you will see a message when the simulation is complete. At this point you may close the window by clicking on the X in the upper right corner of the window.

8. Under the *Results* menu, select *Trajectory Summary*. Here you can view the population growth curve. The time scale is in 1 year time steps. Print the plot of the *Trajectory Summary* by clicking on the printer icon.
9. You can view the numbers corresponding to the plot by clicking the *Show Numbers* button in the upper left corner of the *Trajectory Summary* window. You can return to the plot by clicking the *Show Plot* button, also in the upper left corner of the *Trajectory Summary* window.
10. In Table 3.2, record the *Abundance* of the population at time steps 0, 5, 10, 15, and 20. Your answers should be recorded in the column corresponding to this simulation, which had a carrying capacity (K) of 300. Exit the *Trajectory Summary* by clicking on the X in the upper right corner of the window.
11. Now you are going to run the model using different values for the carrying capacity but keeping the same initial abundance of thirty. Under the *Model* menu select *Population*. Change the carrying capacity to 100. Run the simulation and print your results by repeating steps 8 through 10. As in the previous step, record the average abundance of the population at time steps 0, 5, 10, 15, and 20 in the appropriate column of Table 3.2.
12. Repeat the procedure for a carrying capacity of 50 and again for a carrying capacity of 3000.

Table 3.2. Abundance of osprey over time using the same initial abundance but different values of carrying capacity (*K*).

	Average Abundance			
Time	K = 300	K = 100	K = 50	K = 3000
0				
5				
10				
15				
20				

Questions

1. With a carrying capacity of 300, how long does it take the population to reach 100 individuals?

2. With a carrying capacity of 3000, how long does it take the population to reach 100 individuals?

3. What is the general effect on the population of increasing the carrying capacity under contest competition?

4. Did the population always come close to the carrying capacity (i.e., within two individuals) in your trajectories? Why might it not do so?

5. If the osprey population on Long Island is indeed limited by contest competition over nest sites, how might the manager use this information to help the population recover more rapidly?

Exercise C: Ceiling model of density dependence for barnacles

Background

In this third exercise, you will examine how the ceiling model affects population growth. In the simulation you will be predicting the growth of a barnacle population on a rocky shore. Barnacles are small crustaceans that live in the intertidal zone and feed by filtering the detritus during high tide. Barnacle larvae (immature stages) are planktonic and swim freely in the water, feeding on algae and detritus. Upon reaching the adult stage, individuals attach themselves to a hard surface and construct a hard calcified shell. Once adult barnacles attach to a substrate, they cannot move. If the larvae do not find a place to attach to when they mature, they will die and never be able to reproduce. Imagine that barnacles are introduced to a rocky shore that has no barnacles. The carrying capacity in this model is determined by the available space. All individuals that find a place have an equal chance to survive and reproduce, even in high densities. The population growth rate is likely to remain above 1 until all available space is taken. The population dynamics of the barnacles may mimic the ceiling model, assuming that everything else in the environment stays the same.

Exercises

1. Open a new file in the *Population Growth* (single population models) window of RAMAS EcoLab by clicking on the *New* file icon in the upper left corner of the window. You will be asked if you want to save the file from the previous exercise, but you need not do so.
2. The *General Information* window from the *Model* menu will automatically open. In that window, title this file *Ceiling competition in barnacles*.
3. Also in the *General Information* window, change the *Duration* to 20. By doing this, you have set the program to simulate 20 time steps of reproduction. In this exercise, each time step is 1 year long.
4. Under the *Model* menu select *Population*. Change the *Initial Abundance* to 50. You will assume that 50 barnacle larvae have been introduced into the water and are restricted to the habitat that we are monitoring.
5. Also in the *Population* window, enter a *Growth Rate* of 1.80. This is the growth rate that the population will experience without the effects of density dependence.
6. To investigate the effects of ceiling competition, select *Ceiling* in the pull-down menu titled *Density Dependence Type*. Also in the *Population* window, set the carrying capacity to 100,000. The substrate can hold a maximum of 100,000 adult barnacles. Click OK to exit this window.
7. Now you are going to run the simulation; under the *Simulation* menu select *Run*, or press *Ctrl-R* on your keyboard. At the bottom right corner of this window you will see a message when the simulation is complete. At this point you may close the window by clicking on the X in the upper right corner of the window.
8. Under the *Results* menu, select *Trajectory Summary*. Here you can view the population growth curve. The time scale is in 1 year time steps. Print the plot of the *Trajectory Summary* by clicking on the printer icon.
9. You can view the numbers corresponding to the plot by clicking the *Show Numbers* button in the upper left corner of the *Trajectory Summary* window. You can return to

42 Laboratory 3

the plot by clicking the *Show Plot* button, also in the upper left corner of the *Trajectory Summary* window.

10. In Table 3.3, record the *Abundance* of the population at time steps 0, 5, 10, 15, and 20. Your answers should be recorded in the column corresponding to this simulation, which had a growth rate of 1.80. Exit the *Trajectory Summary* by clicking on the X in the upper right corner of the window.

11. Now you are going to run the model using different values for the growth rate. Under the *Model* menu select *Population*. Change the growth rate to 1.60. Run the simulation and print your results by repeating steps 7 through 10. As in the previous step, record the abundance of the population at time steps 0, 5, 10, 15, and 20 in the appropriate column of Table 3.3.

12. To compare the shape of the population growth curves using ceiling and contest models of competition, you will now run the exercise using contest competition. Under the *Model* menu select *Population*. Select *Ceiling* in the pull-down menu titled *Density Dependence Type*.

13. Change the growth rate back to 1.80. Run the simulation and print your results by repeating steps 7 through 10. As in the previous step, record the abundance of the population at time steps 0, 5, 10, 15, and 20 in the appropriate column of Table 3.3.

14. Repeat the procedure using contest competition at a growth rate of 1.60.

Table 3.3. Abundance of barnacles over time using different models of competition and different growth rates.

	Average Abundance			
	Ceiling		Contest	
Time	$R = 1.80$	$R = 1.60$	$R = 1.80$	$R = 1.60$
0				
5				
10				
15				
20				

Questions

1. How long does it take for the barnacle population to double in size under ceiling competition with a growth rate of 1.80? How long does it take for the population to reach the ceiling under these conditions?

2. How does the time to reach the ceiling change as you change the population growth rate?

3. For a growth rate of 1.80, how long does it take to reach the carrying capacity under contest competition compared to the ceiling model? Why is there a difference?

> *Your lab report should include the following:*
>
> 1. Three trajectory summary graphs, a completed population abundance table (Table 3.1), and answers to questions 1 through 4 for Exercise A
> 2. Four trajectory summary graphs, a completed population abundance table (Table 3.2), and answers to questions 1 through 6 for Exercise B
> 3. Four trajectory summary graphs, a completed population abundance table (Table 3.3), and answers to questions 1 through 3 for Exercise C

References

Akçakaya, H. R., M. A. Burgman, and L. Ginzburg. 1999. *Applied Population Ecology using RAMAS® EcoLab.* Second Edition. Sinauer Associates, Sunderland, MA.

Begon, M., J. L. Harper, and C. R. Townsend. 1996. *Ecology: Individuals, Populations and Communities.* Blackwell Science, Oxford.

Laboratory 4
Wood Storks and Honeyeaters: Estimating Population Characteristics

To understand the future growth and persistence of an endangered species' population, we need to count individuals and monitor the population in the present. How do ecologists measure population size and growth? What do they do with information once they have it?

Introduction

Conservation biologists interested in managing a population of an endangered species need to know whether the population is growing, declining, or stable at the current abundance. Without this information, we cannot make intelligent recommendations that best serve the population of interest. In this laboratory, we will look at how we collect this information as well as how to use more detailed information to make better predictions about future growth.

Estimating Population Abundance

The minimum knowledge needed to predict future population abundance over time is an estimation of a population's current size and growth rate. How do we determine population size in real populations? Ecologists commonly use censuses to monitor populations. You are probably familiar with the term "census" from discussions about the population size of the United States. To census human populations, demographers either visit or send out a questionnaire to all known addresses to determine how many people of each age are at each address. Clearly, taking a census of an animal or plant population cannot be done this way.

The most accurate way to census a population is to identify each individual and note its age. It is sometimes possible to identify all individuals in small or restricted populations; this is true sometimes with endangered species. Complete information helps us understand changes in the size and composition of a population over time. Scientists can obtain this information in a number of ways. Ecologists might capture all the individuals, mark them, and recount them when they are recaptured later. They might also count individuals while moving through the habitat (i.e., ground counts). Alternatively, ecologists may make estimates from flying over populations and counting nesting or den sites. When they try to count all individuals in a population, they may have inaccurate estimates of the population if they assume incorrectly that we have captured or seen all of the members of the population.

To illustrate the effect different census methods can have on an estimate of population size, consider the following example of muskox censusing. The data shown in Table 4.1 are based on estimates of the size of a muskox population on Nunivak Island (Spencer and Lensink 1970). They estimated the population size using two methods, an aerial count and a ground count. As you can see, although they are counting the same population, the two methods provide different estimates of population size.

Table 4.1. Muskox population estimates from Nunivak Island (data from Spencer and Lensink 1970).

Year	Ground Count	Aerial Count
1965	532	514
1966	610	569
1967	700	651
1968	750	714

This muskox example illustrates some of the difficulties in counting individuals. In large populations, it is often not possible to mark or identify all individuals. For these populations, a sample is taken to try to estimate the size of the whole population from the subset of individuals we can count and identify. For sessile (immobile) organisms, the most common method is to count all individuals in several randomly placed quadrats (rectangular areas of a specified size and shape) and multiply the average density within the quadrats by the total area. For mobile organisms one commonly used census method is a mark-recapture survey. A researcher sets up live traps and marks and releases each animal captured. After each trapping bout, he or she compares the number of new animals caught to the number of animals previously caught and marked. If many new animals are captured, it is likely that the individuals originally marked represent only a fraction of the population. If many of the marked animals are recaptured the second time, this is evidence that most of the population has been included in the count. Another way to estimate population size is using a line-transect method. This method involves counting all the individuals in a strip of land and then multiplying the density of individuals in that strip by the total area in which the population is found.

Age and Stage Classes

A population is a diverse collection of individuals. Each individual in a population follows a particular path through life, has a common life history. All individuals of a population share common characteristics, such as the minimum year they can start to reproduce and the maximum life span. At a finer scale, individuals pass roughly at the same time through periods of infancy, adolescence, early adulthood, and old age. Although not all individuals live through all stages of their life, from birth they face a similar life history. Because individuals mature and begin to reproduce at the same age, and have basically the same reproductive life span, individuals of the same age often will be more similar to each other than to individuals of other ages.

When we model a population, we often categorize individuals into age classes. These models are considered **age-structured**. In an age-structured model, the classes are equally sized intervals; a population with classes of 5 year intervals would be divided into 0 to 5 year olds, 5 to 10 year olds, 10 to 15 year olds, and so on. After a time period equal to the interval, all individuals advance to the next class. For example, if the classes are 5 years, all individuals in class 1 will advance to class 2 after 5 years have passed. It is important that all age classes have the same interval, except the oldest class. Individuals in the oldest class will remain in that class until they die, regardless of the time interval.

Alternatively, in some species, reproduction may be based more on body size than on age. Many plants, fish, and reptiles follow this pattern. In these cases it is better to group individuals according to their life history stage and not age. As an example, a young deer may be classified as a juvenile with all other individuals who are older than 1 year of age but not sexually mature. For those species more easily classified by stages, one may use a stage-based modeling approach. Stages may be based on size, reproductive state, or other defining characteristics. Individuals remain in one stage until they acquire the characteristics of the next stage. In the case of the young deer, an individual will remain in the juvenile stage until it reaches sexual maturity and moves into the next stage, adult.

Survival Rate

If we use a structured model, based on ages or stages, we need to know the proportion of animals in each age class to determine the population growth rate. We can determine the survival rate of individuals by counting all individuals in one age class, and then counting them again after an interval of time, when they enter the subsequent age class. If the first count is of individuals of age class 1, in the second count these individuals will have advanced to age class 2. The number we count the second time divided by the number we counted the first time is the survival rate for that transition. Shown below is the equation that describes this calculation.

$$\left(\begin{array}{c}\text{Survival from age class } i \\ \text{to age class } i+1\end{array}\right) = \frac{\text{Number of individuals in age class } i+1 \text{ at time } t+1}{\text{Number of individuals in age class } i \text{ at time } t}$$

$$S_i = \frac{N_{i+1}(t+1)}{N_i(t)}$$

To illustrate a sample estimation of the class-specific survival rates, Table 4.2 is an example of census data from the eastern spotted woodminer, a hypothetical endangered species. We would like to estimate the survival rate for each age class listed in the table. In this example, each age class represents one year. We will assume that all individuals die before they reach 5 years of age.

Table 4.2. Census data for 1993 and 1994 for the woodminer.

Age	1993	1994
0	45	37
1	42	36
2	36	38
3	33	30
4	28	23

In this case the survival of individuals of age class 1 in 1993 that became members of age class 2 in 1994 would be calculated as follows:

$$S_1 = \frac{N_2(1994)}{N_1(1993)} \qquad S_1 = \frac{38}{40} = 0.90$$

To complete this example, the estimated rates of survival for each age class are shown in Table 4.3. Test yourself to see if you can obtain the same values.

Table 4.3. Survival rates for each age class of the woodminer estimated from the data shown in Table 4.2.

Age	Survival Rate
0	0.80
1	0.90
2	0.83
3	0.70
4	0.00

Measuring Fecundity

Population growth requires reproduction. Fecundity is a measure of how many offspring each individual, on average, has in each time step. You compute this by dividing the total number of all offspring produced in a time step by the number of potential parents. Realize that all members of age class 0 in a particular year are those that were born in the previous year. In this example, all individuals in age classes 1 through 4 are reproductively mature and are potential parents. The fecundity value for woodminers in 1993 would be estimated as follows (notice that individuals of age class 0 are not included in the tally of possible parents, because it is assumed they are too young to have offspring):

$$\text{Fecundity of age classes 1 to 4} = \frac{\text{Number in age class 0 in 1994}}{\text{Number of reproductively mature individuals in 1993}}$$

$$F_{1-4} = \frac{(37)}{(42 + 36 + 33 + 28)} = 0.27$$

Of course it is not possible that each individual has 0.27 offspring because each individual produces some integer number of offspring, 0, 1, 2, 3, etc. This fecundity value of 0.27 tells us that approximately 27% of the individuals in age classes one through 4 produced an offspring that year. It is important to note that, especially in populations with low fecundity such as this one, not all individuals are necessarily producing an offspring each year, whereas other individuals may be producing several. Furthermore, if we have very detailed information about the number of offspring produced by individuals of each age class, we can calculate an age- (or stage-) specific fecundity, as we did for survival rate. Otherwise, as in this case, we only have an average fecundity, which we assume to be the fecundity for all classes; this is why the fecundities of all three age classes are the same in Table 4.4.

What Is a Life Table?

What can be done with the data on survival rate and fecundity of the population? One simple way of viewing the data is to present it is as a life table. A life table lists the survival rate and fecundity of each age class in the population. Table 4.4 is a life table constructed of the woodminer population based on the data given in Table 4.2. In this life table, the age classes 3 and 4 have been combined into one age class, ages 3 and above (3+). The survival rate of this new age class can be calculated using the data in Table 4.2 by dividing the number of 4 year olds in 1994 by the sum of the numbers of 3 and 4 year olds in 1993, which gives us 23/(33+28) = 0.38. Remember, this assumes that all 4 year olds in 1993 (28 of them) died before 1994.

Table 4.4. A sample life table constructed for the woodminer population based on the data from 1993 and 1994.

Age	Survival Rate	Fecundity
0	0.80	0
1	0.90	0.27
2	0.83	0.27
3+	0.38	0.27

What Is a Matrix?

From a life table we can predict the future fate of individuals in a population. In the woodminer example, 90% of the individuals in age class 1 are likely survive to become members of age class 2, and on average, each individual in age class 1 will produce 0.27 offspring in the next time step. So you could multiply the number of 1 year old individuals by these values to determine how many will survive to age 2 and how many offspring (age class 0) they will produce. This concept is fairly simple, but it is messy to compute for all age classes. This can be simplified by making a matrix, which puts all these values into a form that is easier to handle, both when doing calculations by hand and in a computer program. The woodminer population's age matrix constructed from the survival and fecundity rates is shown in Table 4.5.

This matrix has four rows and four columns, for a total of 16 numbers (many of which are zeros). The numbers in the matrix represent survival and fecundity values for the population. Above each column and to the left of each row are headers indicating the age class transitions. The first row is for fecundities of each age (or stage) class, which we have calculated to be 0.27 for age classes 1 and above. These numbers represent the average number of offspring per individual that each age class contributes to age class 0. The other numbers in the table represent the survival values. Each survival value indicates the proportion of individuals surviving from the age class in the column to age class in the row. Therefore, the survival rate from age 0 to age 1 (0.80) is placed in the first column, second row. Note that the survival of age class 3+ is specified for a single class. This is because all individuals in age class 3+ remain in class 3+ until they die.

Table 4.5. An age matrix constructed for the woodminer population based on the data from 1993 and 1994.

	From Age 0	From Age 1	From Age 2	From Age 3+
To Age 0	0.00	0.27	0.27	0.27
To Age 1	0.80	0	0	0
To Age 2	0	0.90	0	0
To Age 3+	0	0	0.83	0.38

The information in a stage matrix can be represented as a flow diagram, where each box represents a stage, the arrows represent transitions between stages (i.e., surviving to the next stage or having offspring), and the numbers along the arrows are the survival rates and fecundities. A flow diagram for the woodminer population is shown in Figure 4.1.

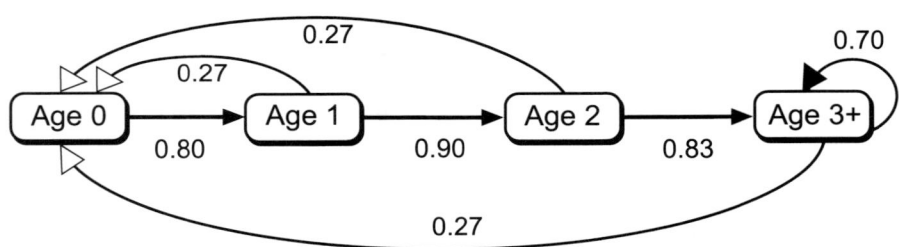

Figure 4.1. Flow chart representing stage transitions for the woodminer population.

In the following exercises, you will look at different census techniques used to monitor wood stork populations in Florida. You will examine the effects these different techniques have on the estimation of population size. You also will be creating a life table from census data for the helmeted honeyeater. Based on census data you will calculate survival rates for each class and average fecundity.

Exercise A: Counting wood storks

Background
Wood stork (*Mycteria americana*; Figure 4.2) populations in Florida have decreased from approximately 75,000 individuals in the 1930s to close to 10,000 in the 1980s (Rodgers et

al. 1995). This dramatic decrease in population numbers resulted in the listing of the wood stork as a federally endangered species in the mid-1980s. The National Audubon Society and the Florida Game and Fresh Water Commission (GFC) have been working together trying to maintain accurate estimates of the stork population.

Wood storks are colonial nesters, meaning many individuals build nests together in a small area. Gaining access to nesting colonies can be difficult, especially in areas with thick ground cover or uncooperative landowners. Aerial surveys have been used as an alternative to ground counts. The agencies monitoring the populations initially assumed that the accuracy of each of the two techniques were comparable. Rodgers et al. (1995) compared the population estimates made from a small aircraft to those on the ground. This type of systematic comparison is rare; often we assume the methods are acceptable and find out too late that we have incompatible and/or inaccurate data.

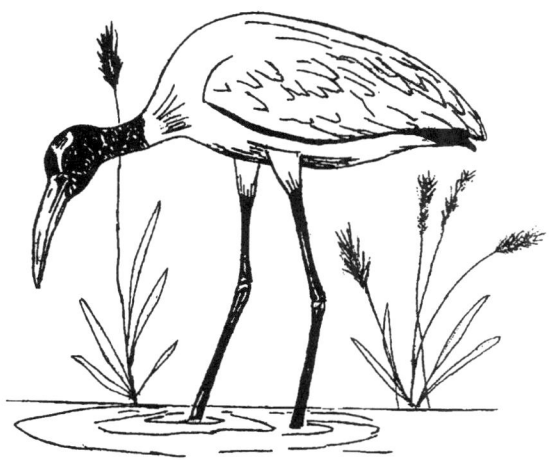

Figure 4.2. The wood stork.

The accuracy of the different techniques for surveying the stork populations may depend on several things. One potential complication is that because storks nest with other shore birds, technicians may sometimes count other white birds as storks or fail to count some of the storks. From the air it may be too difficult to distinguish the storks from other species. Another potential problem, which turns out to be very minor in this case, is that the vegetation coverage over the colonies may interfere with an accurate count of birds. For reasons such as these, ground counts often prove to be more accurate than aerial counts. However, ground counts tend to be more labor-intensive and expensive, so aerial counts are generally more cost-effective.

Exercises

1. Table 4.6 provides population estimates for several populations of wood storks from both aerial and ground counts. Using these data compute the bias and difference for each population estimate using the following equations and fill in Table 4.6.

$$\text{Bias} = \frac{\text{(ground count)}}{\text{(aerial count)}}$$

$$\text{Difference} = |\text{ground count} - \text{aerial count}|$$

(Express difference as an absolute value, i.e., as a positive number)

Table 4.6. Population estimates of several populations of wood storks from both aerial counts and ground counts.

Colony Name	White Birds	Aerial Count	Ground Count	Bias	Difference
Brewster	Low	10	16		
Chaires	Low	225	278		
Dee Dot	High	70	128		
Grand Farm Island	High	15	21		
Lake Yale	High	75	122		
Mulberry	Low	57	57		
Pelican Island	Low	225	160		
Pleasant Grove	High	212	155		

2. Make a graph with the aerial count on the *x*-axis and the difference on the *y*-axis.

3. Make a graph with the ground count on the *x*-axis and the bias on the *y*-axis.

4. Find the average bias for all stork colonies that have a high proportion of other white birds nesting in the stork colony.

5. Find the average bias for stork colonies nesting in areas with a low proportion of other white birds nesting in the colony.

6. Make a bar graph comparing the average biases you calculated for high and low colonies.

Questions

1. Based on the graphs you drew, is there a relationship between the size of the colony and the error (difference) in the population estimate? How about colony size and bias?

2. Look at the bar graph comparing the proportion of other white birds nesting in the stork colonies. Is there more bias in colonies with many other white birds than with few? Why or why not?

3. Do you think there is too great of a disagreement between the aerial and the ground counts of this stork population? How did you decide this?

4. Would you be more or less likely to allow disagreement between census methods in endangered populations than in large stable ones?

Exercise B: The helmeted honeyeater

Background

The helmeted honeyeater (*Lichenostomus melanops cassidix*) is an endangered songbird species endemic to Victoria, Australia. The helmeted honeyeater lives in eucalyptus swamps where individuals establish territories. To monitor the population, it is important to determine the rate at which it is growing. In this example, annual censuses shown in Table 4.7 were taken from a hypothetical swamp population. Individual honeyeaters were counted and placed into the appropriate class based on age. Breeding in this species begins at 1 year of age.

Table 4.7. Hypothetical census data for the helmeted honeyeater.

Age	1991	1992	1993
0	26	28	27
1	16	17	20
2	12	11	13
3	9	8	9
4	7	6	6
5	5	4	5
6	4	3	3
7	3	3	2
8	2	2	2
9	1	1	1
Total	85	83	88

Exercises

Estimating Vital Rates

1. Compute the survival rate for each age class from 1991 to 1992. Repeat for the interval from 1992 to 1993. Fill in Table 4.8.

Table 4.8. Age specific survival rates for the helmeted honeyeater.

	Survival Rate	
Age	1991 to 1992	1992 to 1993
0		
1		
2		
3		
4		
5		
6		
7		
8		

56 Laboratory 4

2. Compute the average fecundity for the population from 1991 to 1992. Refer to the earlier example for help.

3. Compute the average fecundity from 1992 to 1993.

4. Construct a life table (Table 4.9) for the honeyeater from the estimates for 1991 to 1992.

Table 4.9. Life table for the helmeted honeyeater using 1991 to 1992 data.

Age	Survival Rate	Fecundity
0		
1		
2		
3		
4		
5		
6		
7		
8		

5. Using the information from the life table, construct an age matrix (Table 4.10). All age classes over 3 years have been combined to make an "older adult" class. The survival rate for age-class 3+ has been given to you, because this value must be calculated over several age classes.

Table 4.10. Age matrix for the helmeted honeyeater using 1991 to 1992 data.

	From Age 0	From Age1	From Age 2	From Age 3+
To Age 0				
To Age 1				
To Age 2				
To Age 3+				0.71

Questions

1. Were the values for the survival rates and fecundities the same for both 1991 and 1992? Which of the following would represent the population's survival rates best: those for 1991, those for 1992, or the average of the two? Why?

2. We have 3 years of census data. Are these data sufficient to represent the population? What are some reasons we may want more data?

An Age-Structured Model

You are now going to use the matrix to run a simulation with an age-structured population. You will be using RAMAS EcoLab, the same program you used to look at population growth in unstructured populations.

1. Open the program RAMAS EcoLab (double-click on its icon with your mouse).
2. Click on *Age and Stage Structure*.
3. Go to the *File* menu. Select *New*.
4. Under the *Model* menu select *General Information*. In that window title this file *Helmeted Honeyeaters*.
5. Also in the *General Information* window, change the *Duration* from 0 to 20. By doing this, you have set the program to simulate 20 time steps of reproduction. In this exercise, each time step is 1 year long.

6. Also in the *General Information* window, set the number of *Replications* to 1. You will run one simulation to view how the size of the helmeted honeyeater population is expected to change over time. Click OK to exit this window.
7. Under the *Model* menu select *Stages*. Click the Add button until you have 4 stages. Under *Name* enter the age classes found in the age matrix (Table 4.10). Make no changes to the *Average Weight* column. Click OK to exit this window.
8. Under the *Model* menu select *Initial Abundances*. The age classes you entered in the previous step will appear in the window. Under each age class, enter the appropriate number of individuals from the 1993 census (Table 4.7). Remember to combine the number of individuals in age classes 3 through 9, and enter this value for the initial abundance of age class 3+.
9. Under the *Model* menu select *Stage Matrix*. Enter the fecundities and survival rates as they are listed in the life table (Table 4.10). Click OK to exit this window.
10. Now you are going to run the simulation; under the *Simulation* menu select *Run*, or press *Ctrl-R* on your keyboard. At the bottom right corner of this window you will see a message when the simulation is complete. At this point you may close the window by clicking on the X in the upper right corner of the window.
11. Under the *Results* menu select *Trajectory Summary*. Here you can view the population growth curve. The time scale, like the age classes, is in 1 year time steps. Print the plot of the *Trajectory Summary* by clicking on the printer icon.
12. You can view the numbers corresponding to the plot by clicking the *Show Numbers* button in the upper left corner of the *Trajectory Summary* window. You can return to the plot by clicking the *Show Plot* button also in the upper left corner of the *Trajectory Summary* window.
13. In Table 4.11, record the initial and final *Average Abundance* of the population (the population sizes at time steps 0 and 20) in the column labeled 1991 to 1992. Exit the *Trajectory Summary* by clicking the X in the upper right corner of the window.

Table 4.11. Initial and final abundances for the 1991 to 1992 and the 1992 to 1993 survival and fecundity rates.

	1991 to 1992	1992 to 1993
Initial Abundance		
Final Abundance		

Now you will rerun the simulation with the fecundity and survival rate values for 1992 to 1993. The initial abundance will remain the same.

14. Under the *Model* menu, select *Stage Matrix*. Enter the fecundities and survival rates you calculated in the previous exercise. The fecundity of age classes 1 through 9 for 1992 to 1993 was calculated in step 3, and the survival rates were calculated in Table 4.8. Use a survival rate of 0.70 for age class 3+. Click OK to exit this window.
15. Now you are going to run the simulation using these new values of survival and fecundity. Under the *Simulation*, menu select *Run*, or press *Ctrl-R* on your keyboard. Close the window by clicking on the X in the upper right corner of the window.

16. Under the *Results* menu, select *Trajectory Summary*. Here you can view the population growth curve. The time scale, like the age classes, is in 1 year time steps.
17. You can view the numbers corresponding to the plot by clicking the *Show Numbers* button in the upper left corner of the *Trajectory Summary* window. You can return to the plot by clicking the *Show Plot* button also in the upper left corner of the *Trajectory Summary* window.
18. In Table 4.11, record the initial and final *Average Abundance* of the population (the population sizes at time steps 0 and 20) in the column labeled 1992 to 1993. Exit the *Trajectory Summary* window by clicking on the X in the upper right corner.

Questions

1. What does the graph look like with the 1991 to 1992 survival and fecundity values? Is the population growing?

2. Is the population growing under the 1992 to 1993 survival and fecundity rates? How does this compare to the population growth under the 1991 to 1992 rates?

3. If you knew that the 3 years of data we have are from years with good weather and a lot of food, what do you think the population growth rate may look like for a bad year? What are some ecological or environmental conditions that would cause this population to have a bad year and decline?

4. When you changed both fecundity and survival rates to reflect the data between 1992 and 1993, how much higher did the population size end up than when you used the rates from 1991 to 1992? What does this imply about the accuracy of measured fecundity and survival rates?

Your lab report should include the following:

1. Completed Table 4.6, 3 graphs, answers to steps 4 and 5, and answers for questions 1 through 4 for Exercise A
2. A survival rates table (Table 4.8), answers to steps 2 and 3, a life table (Table 4.9), an age matrix (Table 4.10), and answers to questions 1 and 2 for Part 1 of Exercise B
3. Two trajectory summary graphs, completed average abundance table (Table 4.11), and answers to questions 1 through 4 for Part 2 of Exercise B

References

Rodgers, J. A., S. B. Linda, and S. A. Nesbitt. 1995. Comparing aerial estimates with ground counts of nests in wood stork colonies. *Journal of Wildlife Management*, 59(4): 656–666.

Spencer, D. L. and C. J. Lesink. 1970. The muskox of Nunivak Island, Alaska. *Journal of Wildlife Management*, 34: 1–15.

Laboratory 5
Grizzly Bears: The Problems of Small Populations

Small and rare populations often receive more attention from ecologists than large populations. This laboratory explores the reasons for and implications of high extinction risk in small populations.

Introduction

Changes in the size of a population over time can have a wide range of consequences depending on the species, the location, and many other factors. For example, unlimited growth of the white-tailed deer has caused widespread habitat destruction in the upper Midwest parks. On the other hand, the number of Florida panthers has declined to the point that many conservation biologists are skeptical that the species can avoid extinction. As you explored in the previous lab, estimating the population growth rate, although sometimes difficult, can be quite useful for making predictions about future abundance. The ability to predict future abundance is a critical part of developing long-term management plans for an area of land or a species.

Environmental and Demographic Variation

All populations experience some variation in their numbers over time. Some of this variation is predictable. If a population has a distinct breeding season, each year after the females produce offspring, the population will grow in size. If there is a season that is particularly hard on a population, the population abundance will fall during this time. However, some variation is not predictable. We cannot forecast all the variation in abundance we will see in a population. This random, unpredictable variation is called **stochasticity**. Two types of random variation a population may experience are demographic and environmental stochasticity. **Demographic stochasticity** is random variation in sex ratios, birth rates, and death rates. Populations are made up of individuals, which are not exactly alike. As a result, often more or fewer individuals die or are born than would be expected. Demographic variation occurs within a population even if the environment does not vary. If females in a population with demographic stochasticity produce an average of two offspring per year, some females may have no offspring, some may have one offspring, and some may have two, three, or four offspring. This variation is random.

Here is an example of demographic stochasticity in a group of mountain gorillas (*Gorilla beringei*). Females breed every 3.5 to 4.5 years (average fecundity rate of 0.25). If three of four females in a group give birth during the same year, and the fourth female gives birth the following year, there may be 3 years before the next offspring is born. If the average fecundity of the females is 0.4 and there are four females in the group, we would expect 1 offspring a year (0.25 fecundity per year × 4 individuals = 1 offspring per year). However, we can have multiple births in 1 year followed by several years with no births. Likewise, several individuals may die during the same year due to chance. Having more infants born or more individuals die than expected will change the average growth rate. If a population is very small, demographic variation can have very large consequences.

Unpredictable variation in the environment, or **environmental stochasticity**, also creates variation in population growth rate over time. Although some environmental variation is predictable, like the changing seasons, variation between different years tends to be less predictable. Rainfall, temperatures, storms, humidity, and amount of cloud cover can all vary between years. Weather conditions directly affect the amount of food available each year. Some years may be very good for a species, whereas other years may have food shortages, drought, extreme temperatures, or other adverse conditions. During bad years birth rates may drop and death rates may increase. Taken together, low birth rates and high death rates can substantially decrease the population size.

In a large population with a wide distribution, stochastic effects generally are not too harmful. However, if a population is very small, or is found only in a small area, demographic and environmental stochasticity can drive a population to the brink of extinction. A large population is usually able to recover from bad years or catastrophes because some individuals should survive to reproduce in future years. If 90% of a population of 3,000 is killed during a catastrophe, 300 individuals would still remain to continue to reproduce. However, if the same catastrophe kills 90% of a population of 30 individuals, only 3 individuals would remain. If two are male and one is an old, nonreproducing female, the population will become extinct.

Loss of Genetic Variation in a Population

Another problem faced by small populations is the loss of genetic variation at both the individual and population level. All diploid organisms (most multi-cellular organisms are diploid) are defined as having two copies of each gene called alleles. One copy is inherited from each parent. Although each allele contains information about the same trait, they may not be identical. When an individual has corresponding alleles that are different, it possesses genetic variation. Having two copies of a gene is beneficial if one copy has some problems, or mistakes. Every organism has at least a few alleles that do not function or might be harmful, but because there are two alleles for each gene, there is a good chance that one will function properly and compensate for the other. Consider an organism with a gene for resistance to a certain disease. If an individual has one nonfunctional allele and another that properly codes for disease resistance, it will not be susceptible to that disease. However, if an individual has two alleles of the same gene that are harmful, there will be a detrimental effect on the individual's fitness.

Populations also possess genetic variation. This variation occurs when individuals in the same population have differences in their genetic makeup. Genetic variation helps populations survive and adapt to changing environments. One gene may be beneficial in one situation, while another may be better in a second situation. If a population is genetically diverse, it is likely that some individuals will have genes necessary to adapt to a change in their environment. The population as a whole then, will have a lower risk of extinction. A population with no genetic variation is less likely to survive environmental change because all individuals may be equally susceptible. For example, geneticists fear that the lack of variation in cheetah populations makes them more vulnerable to introduced diseases such as feline leukemia.

Small populations lose genetic variation over time. This loss of variation is usually due to inbreeding, or breeding with relatives. If a population is small enough, there are not many mates available from which to choose. Eventually, individuals mate with others that are closely related. The more closely related individuals are, the more genes they share. If individuals with the same genes mate, the resulting offspring may inherit two identical alleles. The individual will be less likely to survive, if both of the alleles are inferior. As a population becomes smaller and smaller, it becomes more likely that individuals will inherit two identical copies of a particular gene. In very inbred populations, many individuals have identical copies of harmful genes, often resulting in reduced fertility and unhealthy offspring. For example, evidence has been found of reduced sperm fertility of inbred lions and cheetahs. The problems associated with low genetic diversity due to inbreeding are called inbreeding depression.

The extinction risk of a small population, therefore, may be affected by the genetic health of the population. To protect the long-term viability of a population, we need to ensure that there are enough individuals to protect it from the loss of genetic variation. If a population is below the threshold necessary to maintain genetic variation, the fitness of the individuals in the already small population may decrease. With lower fecundities and/or higher mortalities, it will be increasingly difficult for the small population to recover.

To maximize the amount of variation maintained in a population, zoos and conservation organizations have organized gene banks. Managers of both reserves and zoos keep track of the relatives of all individuals and they plan matings between distantly related individuals. Sometimes this entails collecting sperm from wild animals to inseminate captive females; sometimes it entails shipping individuals between zoos to create the best pair of animals.

Extinction Vortexes and Allee Effects

Another factor that may push populations toward extinction is the **Allee effect**. At very low population densities, individuals may not have enough contact with other individuals of the population. It can be difficult for individuals to find mating partners, causing the reproductive rate and the population growth rate to drop. Although evidence for the Allee effect is limited, it is likely to affect species where the individuals live at low densities, such as ocean-dwelling animals like tuna, swordfish, and whales. Plants that depend on pollinators to visit several individuals of the same species may also be subject to Allee effects.

Once a population becomes too small to sustain itself, it may enter an **extinction vortex**. An extinction vortex is very much like a whirlpool or a black hole; once you begin to be pulled toward its center, it becomes harder and harder to escape. For populations the center of the vortex is extinction, or 0 population size. Populations at very low abundance may not be able to escape the pull toward extinction.

As a population falls below a critical threshold, it becomes too sensitive to environmental and demographic stochasticity to maintain stable numbers. The added affect of inbreeding depression and allee effects may cause its numbers dwindle until the population becomes extinct. Many small and isolated populations may already be experiencing extinction vortexes. Populations in this downward spiral cannot change its course without outside interference and management.

How large must a population be to sustain itself and avoid the problems faced by small populations? This is a difficult question with no clear answers but probably varies somewhat for each species. Some scientists speculate that populations should have at least 1000 to 10,000 individuals to persist long term; others estimate that populations should really have at least 100,000 individuals or more to persist indefinitely (Gilpin and Soulé 1986).

Population Viability Analysis

In the previous sections we discussed some of the characteristics that make a population vulnerable to extinction. If a population is at risk, it is important to look more closely at the likelihood of the population facing extinction. One way to determine the long-term sustainability of a population is to perform a **population viability analysis** (PVA). A population viability analysis is a model used to predict the probability of a population declining or becoming extinct. If a population is not likely to survive in its present condition, PVA can be used to determine which aspects of the population will be the most sensitive to conservation efforts. Then we can explore methods to help increase the population size, the growth rate, or the habitat quality to improve the future probability of survival.

One of the most important estimates we try to predict with a population viability analysis is a **minimum viable population** (MVP) size, or the minimum number of individuals required for long-term survival (Gilpin and Soulé 1986). If the population size falls below the MVP size, it may fall into an extinction vortex. Conservation biologists are very concerned about estimating the MVP size. If a population is currently below the MVP, it will probably be necessary to interfere with current processes to try to increase the population size. If the populations are left alone, below their minimum population size, there is a considerable risk that the population will become extinct.

In a population viability analysis, many simulations are run to predict the future growth rate of the population. After completing these simulations, we compute an average final abundance (the total of the final abundance in all trials/number of trials). We also compute a **standard deviation**, or a measure of the variability between the simulation results. A large standard deviation indicates that the results from each simulation were different, or variable. A small standard deviation means the results from each simulation were fairly similar. If the standard deviation is small, one may have more confidence that the average final abundance reflects what will happen to the population. The standard deviation is shown in the trajectory summary graph as error bars. You will notice that over time the predictions are less precise and the error bars are much wider after 50 time steps than they are at 5 time steps.

Grizzly bear populations

Background

Since 1975, the grizzly bear (*Ursus horribilis*) has been listed as a threatened species in the United States under the Endangered Species Act. As of early 1998, fewer than 1000 individuals survive in the continental United States. They are dispersed in small populations in Montana, Idaho, Wyoming, and Washington. The map in Figure 5.1 shows the areas where grizzly bear populations are currently located (U.S. Fish and Wildlife Service 1999). Fortunately the persistence of the grizzly bear is not dependent on the populations in the continental United States; much larger populations of grizzly bears are present in Alaska and Canada.

As you can see, these grizzly bear populations are not connected by continuous habitat. Between each of these populations are farmland, towns, and cities, making movement of individuals between them very unlikely. Each population is small and, as we have learned earlier in this chapter, small populations are vulnerable to extinction. In the following exercises, you are going to analyze 3 populations of grizzly bears: Yellowstone, Northern Continental Divide, and Cabinet-Yaak ecosystems. You will perform population viability analyses to determine the effect initial population size has on population viability.

Figure 5.1. Map of the existing grizzly bear populations in the Northwest United States (after United States Fish and Wildlife Service Web page)

Exercise A. Viability of several grizzly bear populations

You will first look at the viability of a population of grizzly bears in Yellowstone National Park when stochasticity is not included.

1. Open the program RAMAS EcoLab (double-click on its icon with your mouse).
2. Click on *Age and Stage Structure*.
3. Under the *File* menu, select *Open* and choose *YellowstoneGrizzly.st*. This file contains an age-structured matrix of the grizzly bear population based on data from the United States Fish and Wildlife Service. The age classes are 3 years long.
4. Under the *Model* menu select *General Information*. Notice that the *Duration* is set at 20. Which means that the program is set to simulate 20 time steps of reproduction. In this simulation each time step is also 3 years long.
5. Also in the *General Information* window, notice that the *Number of Replications* is set at 1. You will run one simulation to view how the size of the Yellowstone grizzly population is expected to change over time. Make sure that the *Use demographic stochasticity* box is unchecked. Click *OK* to exit this window.
6. Under the *Model* menu select *Initial Abundances*. Notice the structure of the age distribution. The youngest age classes (Age 0–2) has many more individuals than the other classes because not all individuals survive to maturity. Click *Cancel* to exit this window.
7. Under the *Model* menu select *Stage Matrix*. Inspect the matrix, and note that the fecundities for middle age classes are higher than for young adults or older adults. Click *Cancel* to exit this window.
8. Now you are going to run the simulation; under the *Simulation* menu select *Run*, or press *Ctrl-R* on your keyboard. At the bottom right corner of this window you will see a message when the simulation is complete. At this point you may close the window by clicking on the X in the upper right corner of the window.

9. Under the *Results* menu, select *Trajectory Summary*. Here you can view the population growth curve. The time scale, like the age classes, is in 3-year time steps. Print the plot of the *Trajectory Summary* by clicking on the printer icon.
10. You can view the numbers corresponding to the plot by clicking the *Show Numbers* button in the upper left corner of the *Trajectory Summary* window. You can return to the plot by clicking the *Show Plot* button, also in the upper left corner of the *Trajectory Summary* window. Exit the *Trajectory Summary* by clicking on the X in the upper right corner of the window.

You will now see how demographic stochasticity affects the population growth and viability over time.

11. Open *General Information* from the *Model* menu. Reselect (click in the box) *Use demographic stochasticity*. Click OK.
12. Now you are going to run the simulation; under the *Simulation* menu select *Run*, or press Ctrl-R on your keyboard. At the bottom right corner of this window you will see a message when the simulation is complete. At this point you may close the window by clicking on the X in the upper right corner of the window.
13. Under the *Results* menu, select *Trajectory Summary*. Here you can view the population growth curve. The time scale, like the age classes, is in 3 year time steps. Print the plot of the *Trajectory Summary* by clicking on the printer icon.
14. You can view the numbers corresponding to the plot by clicking the *Show Numbers* button in the upper left corner of the *Trajectory Summary* window. You can return to the plot by clicking the *Show Plot* button, also in the upper left corner of the *Trajectory Summary* window. Exit the *Trajectory Summary* by clicking on the X in the upper right corner of the window.
15. Look through the table of results and find the highest and lowest values from the average abundance column for the entire summary. Note that because we only ran 1 replication, the minimum and maximum for each time step is the same. Record the highest and lowest value for *all* time steps in Table 5.1. Repeat the simulation a total of 10 times, recording the highest and lowest value from the average abundance column for each simulation.

Now you will look at the effect of both initial population size and demographic stochasticity on the model. You are going to look at 2 other populations of grizzly bear, the Northern Continental Divide Ecosystem and the Cabinet-Yaak Ecosystem.

16. Open the file for the *NCDEGrizzly.st*. Repeat steps 11 to 15. Fill in Table 5.2 for this population.
17. Open the file for *Cab-YaakGrizzly.st* (a third grizzly bear population). Repeat steps 11 to 15 and fill in Table 5.3 for this population.
18. Total the number of simulations where the population fell to 0, below 5, and below 20 for each of the 3 populations. Record your values in Table 5.4.

68 Laboratory 5

19. Divide that number by the total simulations (ten) to calculate the probability of each population falling to a given level. Record your values in Table 5.4.

Table 5.1. Results of the Yellowstone Grizzly Population

Simulation	Average Abundance	
	Lowest Value	Highest Value
1		
2		
3		
4		
5		
6		
7		
8		
9		
10		

Table 5.2. Results of the North Continental Divide (NCDE) grizzly population.

Simulation	Average Abundance	
	Lowest Value	Highest Value
1		
2		
3		
4		
5		
6		
7		
8		
9		
10		

Table 5.3. Results from the Cabinet-Yaak ecosystem population.

Simulation	Average Abundance	
	Lowest Value	Highest Value
1		
2		
3		
4		
5		
6		
7		
8		
9		
10		

Table 5.4. Summary of results for all populations.

Population	Number of Simulations with Abundance of:			Proportion of Simulations with Abundance of:		
	0	≤ 5	≤ 20	0	≤ 5	≤ 20
Yellowstone						
NCDE						
Cabinet-Yaak						

Questions

1. What effect does the initial population size have on the probability that a population will fall below a critical level?

70   Laboratory 5

2. In this model, the population is not increasing without demographic stochasticity. If the population were growing, would demographic stochasticity still be important?

3. The simulations you ran only take into account demographic stochasticity. What are some kinds of environmental conditions that may affect the population? Think about the weather and environment where grizzly bears live.

4. Can you minimize either demographic or environmental stochasticity? How? Think of different management options to help keep these populations stable.

Exercise B: Running multiple simulations

1. From the *File* menu, select *Open*. From the available choices, select *YellowstoneGrizzly.st*.
2. Select *General Information* from the *Model* menu. Change the number of replications to 50. Make sure there is a checkmark in the box for *Use Demographic Stochasticity*.
3. Now you are going to run the simulation; under the *Simulation* menu select *Run*, or press *Ctrl-R* on your keyboard. At the bottom right corner of this window you will see a message when the simulation is complete. At this point you may close the window by clicking on the X in the upper right corner of the window.
4. Under the *Results* menu, select *Trajectory Summary*. You have just run multiple simulations together. The program provides a summary of all the simulations. The single line you see on the screen represents the average value over all the simulations at each time step. The vertical bars indicate the amount of variation between outcomes in each of the different simulations. The red symbols indicate the highest and lowest values for each of the simulations.
5. You can view the numbers corresponding to the plot by clicking the *Show Numbers* button in the upper left corner of the *Trajectory Summary* window. You can return to the plot by clicking the *Show Plot* button also in the upper left corner of the *Trajectory Summary* window. The *+1 S.D and −1 S.D.* columns indicate the standard deviation on each side of the average (refer to the introduction for an explanation of standard deviation). Exit the *Trajectory Summary* by clicking on the X in the upper right corner of the window.

Questions

1. In the first two exercises you ran single replications of each simulation and recorded the outcome. What additional information does this give you that is unavailable when you run multiple replications together?

2. If you are losing information by running multiple replications together, what advantage does running many replications give you? Do you have more confidence in the average abundance when you have 10 replications or when you have fifty?

Your lab report should include the following:

1. Two trajectory summaries from Exercise A
2. Table 5.1 through 5.4, and questions 1 through 4 from Exercise A
3. Questions 1 and 2 from Exercise B

References

Gilpin, M. E., and M. E. Soulé. 1986. Minimum viable populations: Processes of species extinctions. In M. E. Soulé (ed.), *Conservation Biology: the Science of Scarcity and Diversity*, pp. 19–34. Sinauer Associates, Sunderland, MA.

Miller, G. T. 1982. *Living in the Environment*. Wadsworth, Belmont, CA.

Schaller, G. 1972. *The Serengetti Lion*. University of Chicago Press, Chicago, IL.

United States Fish and Wildlife Service Grizzly Bear Reintroduction Program: 1999 http://www.r6.fws.gov/endspp/grizzly

Laboratory 6
Giant Pandas: Risks Faced by Endangered Species

Endangered species are found often only in several small isolated populations. As we learned in the last laboratory, small populations are very vulnerable to extinction. When only small populations of a species are left, we need to be doubly concerned about the survival of each individual population. In this laboratory, we will look at risks faced by populations of endangered species.

Introduction

A major focus of conservation biology is the protection of endangered species. Endangered species are species whose population size or the rates of population decline indicate these species will probably become extinct without a reversal of current trends. Many more species are considered threatened or vulnerable, meaning they must be closely monitored because a safe future is not guaranteed. Conservation projects are often attempts to save species that are already in much danger. Dealing with problem species after they have fallen to critical levels, rather than preserving ecosystems or habitats, has been called a "fire-brigade" approach because it is reaction to an existing crisis, not the prevention of further crises. It would be much better to prevent species from becoming endangered in the first place, but conservation biology is fighting an uphill battle; so many species are currently at risk there is little time to implement preventative measures. Conservation biology works much like the repair of a leaky dam. As more leaks appear, the dam as a whole becomes weaker. The best plan is to build a new dam, but often the focus is on repairing leaks to prevent an imminent disaster and there is little time to build a new dam.

Indicators, Flagships, and Keystone Species

Many conservation projects focus on the protection of one or two species in a target habitat. This approach is called a "single species" approach. In a single species approach, conservation biologists focus on the continued survival of an endangered species. The population size, growth rate, resource availability, habitat integrity, toxin levels, and dispersal rates are monitored to determine what factors are responsible for the decline. Management efforts, policy initiatives, captive breeding, and other techniques are then implemented to improve the future of the target population. There are several justifications for using this approach.

One reason for focusing on single species is that their declining population may be indicating that something is wrong or has changed in their habitat. These indicator species can provide a good measure of the overall habitat quality because they are sensitive to changes in the environment. If the indicator species are stable and show no symptoms of population decline, the habitat will probably also be adequate for less sensitive species. Amphibians are often used as indicators because they are particularly sensitive to toxins in the environment and have both aquatic and terrestrial life stages. Amphibian declines have been linked to acid rain, pollution, habitat fragmentation, and ozone depletion. Species with very specific habitat or resource requirements can also act as indicators. If their population abundance is falling, it is probably because of a decrease in the quality or availability of specific resources or habitats. We can also use species as indicators if they reveal changes in the community through changes in diet, ranging patterns, or population density, even if they are not affected as strongly as other species that may be more difficult to monitor. An example of this type of indicator is a predator that has a diet that is sensitive to changes in the availability of prey.

Another reason for focusing on specific species is that some are particularly integral to the persistence of the community in which they live. These important community members are called keystone species. One example might be a species that many other species rely on for food or other resource. Top predators may also function as keystone species by reducing the density of common prey species and allowing more species to coexist. If a keystone predator is removed from a community, the prey community may lose its stability and rare species may be lost from the community.

The sea otter is an example of a keystone species in the northern Pacific. Sea otters eat sea urchins, which in turn eat kelp or seaweed. During the early twentieth century the population of sea otters along the northwest coast of the United States crashed. Once they were gone, the sea urchin population increased. The large numbers of sea urchins destroyed kelp forests. In response to this increase in prey, the sea otter population made a strong recovery along the northern California coastline; and the kelp forests returned as well as the numbers of sea urchins declined. Now the sea otter is facing another crisis. Heavy fishing in Alaskan and northern Pacific waters has caused a decline in the population density of many smaller schooling fish that sea lions depend on for food. The sea lion population has declined drastically in the last two decades. Sea lions are a major prey of the killer whales. With fewer sea lions to support them, killer whales are now preying more intensively on sea otters. Sea otters have been reduced much lower densities than they were a decade ago. As the number of sea otters declines, the number of sea urchins are increases, and the kelp forests have started to disappear again. The absence of the sea otter or the sea lion in these communities changes the structure and balance of many other species present. Conservation biologists believe it is important to identify the most important members in a community and ensure their survival.

The single species approach may be also useful if the species that is protected requires a large area and a wide range of resources to survive. Such species are called umbrella species because protecting them conserves habitat and resources for the many other species that also use these areas and resources. Examples of umbrella species include the wolves of Yellowstone National Park and the marbled murrelet of California. Both of these species exist at low densities and require large areas of suitable habitat to sustain their populations. By preserving the large areas these two species require, many other species that live in the same habitat will also be protected.

A fourth reason for placing conservation focus on a single species is to attract support from the public and funding organizations. Humans are much more sympathetic toward large-bodied species, or those considered attractive or cute. It is easier to mobilize popular support behind a panda bear than a rare beetle or snake. These loveable and marketable species are called flagship species because they can motivate public interest toward supporting the protection of habitats. Some of the flagship species you may have heard about are the mountain gorilla, spotted owl, and the bald eagle.

Endangered Species

The International Union for the Conservation of Nature (IUCN) has developed criteria to assess the extinction risk of different species (IUCN Species Survival Commission 1994). They created 5 major classes: critically endangered, endangered, vulnerable and lower risk. Critically endangered species face an extremely high risk of extinction in the very near future. Endangered species also face a high risk of extinction if current trends are not reversed, although the risk is not as great or as imminent as for critically endangered species. Vulnerable species have reduced abundances and are at risk of becoming endangered. Species classified as lower risk do not meet the criteria of the other three categories and have populations that appear secure for the long-term future.

The United States Fish and Wildlife Service has also developed a classification system to identify species in need of protection based on the Endangered Species Act of

1973 (ESA). In this system, species at risk are classified as either endangered or threatened. Endangered species are in danger of becoming extinct in all or in a significant part of their range, and threatened species are likely to become endangered if conservation measures are not taken. The ESA requires that the populations and habitat of endangered and threatened species receive protection by federal agencies.

Life History and Population Size

There are a number of characteristics that may make species more vulnerable to extinction. Some of these characteristics include low densities, large body size, and specific habitat requirements. Examples of endangered species and the characteristics that make them particularly vulnerable to extinction are summarized in Table 6.1.

Table 6.1. Ecological characteristics of rare or extinction prone species (after Miller 1982).

Characteristic	Examples
Large body size	Elephant, rhino, blue whale, grizzly bear
Low reproductive rate	Mountain gorilla, Andean condor, whooping crane
Specialized requirements	Spotted owl, marbled murrelet, black footed ferret
Low population density	Bengal tiger, African wild dog, African lion, jaguar
Limited distribution	Lemur, Komodo dragon, pygmy hippopotamus, Grevy's zebra

Some species are always found in small numbers or at very low densities. These species are considered rare. Some species are naturally rare. Rarity by itself is not necessarily a bad trait, but if rarity is combined with a loss of available habitat, the population may decline to dangerously small numbers. If a species consists of only a few small populations, any large stochastic change can cause an entire population to become extinct and bring the species as a whole closer to extinction.

Large-bodied, long-lived animals are usually found in lower population densities than small-bodied animals. The size of a protected area, such as a national park or a nature reserve, may limit the population size of large-bodied species. These species need large areas to support a viable population and are more vulnerable to extinction if large areas of their habitat are destroyed. In small parks, we may believe we have protected populations of rare or endangered species satisfactorily. These small populations may be able to survive for short periods of time, but because of their size, they are at high risk of extinction over longer time periods. An important point to remember is an area that may support a viable population for one species may not be adequate for other species. Some species will need very large areas to be able to survive in the long term. Figure 6.1 shows the estimated population size of spotted hyenas in different-sized reserves in East Africa (Schaller 1972). Notice how much larger population sizes are in the bigger parks. As the population size increases, usually the risk of extinction decreases. It is likely that populations of hyenas in larger parks will face fewer extinction events.

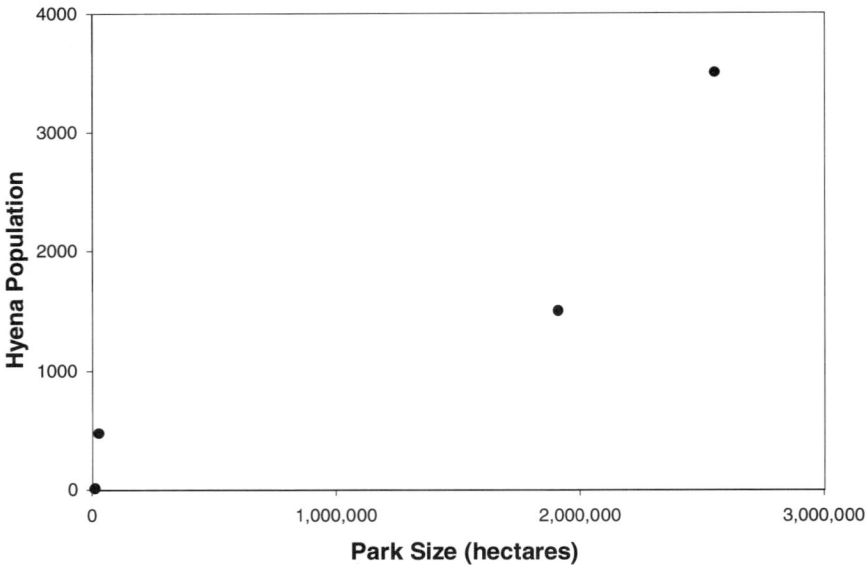

Figure 6.1. Population sizes for spotted hyenas in reserves in East Africa (based on Schaller 1972).

Ecological traits and natural history also have a large effect on the densities and locations where species occur. Some species are found in a very small geographic area. Many island species are **endemic**, meaning they only occur on a single island or in a very restricted area. Species that have very specific habitat requirements can only be as abundant as their resources. Many rare birds, such as the spotted owl, the marbled murrelet, and the red-cockaded woodpecker, are limited by the availability of suitable nesting locations. Predators also tend to occur at low densities because they are limited by the amount of prey available. Individual predators tend to need large home ranges to support themselves. African wild dogs, lions, wolves, tigers, and hyenas all occur at very low population densities. The Florida panther represents a good example of the problems of species with large home ranges and low population density. As the amount of protected area in southern Florida decreases, panthers are forced to cross roads and travel much greater distances for food and mates. Mortality has dramatically increased in the panthers as a result of these changes.

Threats to Endangered Species

Different circumstances are often responsible for the creation of endangered species. The major reasons species become endangered are habitat loss and fragmentation, overexploitation, and the introduction of diseases or new species into a sensitive community. Table 6.2 illustrates the number of animal species that have become vulnerable or extinct due to different external factors.

Overexploitation from hunting or harvesting often has a major impact on populations of rare species. Often illegal harvest, or poaching, continues to deplete populations after they have been identified as vulnerable or endangered. As many as 30% of threatened species are affected by overexploitation or illegal harvest. For some species, such as primates in West Africa, poaching is a more serious immediate threat than habitat loss. Many fisheries continue to decline because it is not possible to regulate harvesting in international waters.

Table 6.2. Factors responsible for extinction or risk of extinction in a number of animal groups. (Based on data from Reid and Miller 1989.)

Group	Number of Species Lost Due to Each Cause					
	Habitat Loss	Over-exploitation	Species Introduction	Predators	Other	Unknown
EXTINCT						
Mammals	19	23	20	1	1	36
Birds	20	11	22	0	2	37
Reptiles	5	32	42	0	0	21
Fishes	35	4	30	0	4	48
THREATENED OR ENDANGERED						
Mammals	68	54	6	8	12	-
Birds	58	30	28	1	1	-
Reptiles	53	63	17	3	6	-
Amphibians	77	29	14	-	3	-
Fishes	78	12	28	-	2	-

Human population growth and resource consumption are placing increasing pressure on the environment. It often is difficult to perceive the damage inflicted on the environment at the individual level. It is only when we step back to see the effects of many people that the declines in species populations and fragmentation of habitats are obvious. Species geographic ranges become smaller and smaller as they lose more and more of their habitat to agriculture, development, and resource harvesting. Hunting and exploitation of natural resources increase as previously isolated regions become more accessible through road-building, logging, and hunting.

An unintentional by-product of human development is the introduction of new species into sensitive environments. Sometimes these species are competitively superior to the native species or they may be very effective predators. New species may also come in the form of pathogens. Native species that have not evolved resistance may be susceptible. In each of these cases, the introduction of new species has a detrimental effect on the native populations and communities.

International regulation often is necessary to monitor the stability and health of species populations. International markets drive much of the harvesting of resources and animals; those who buy international goods may not understand the impact trade has on species and habitats. For this reason, there is an international regulatory policy to monitor the impacts trade is having on species. This treaty is called CITES, or the Convention on International Trade in Endangered Species. CITES has established lists of species considered endangered and vulnerable (based on criteria similar to IUCN); trade and exploitation of these species are prohibited or closely regulated.

Giant Pandas in China

Background

Giant pandas are found in the mountains of China. The range and population size of the panda population has decreased dramatically due to habitat fragmentation and deforestation. Figure 6.2 illustrates the present panda distribution as well as the fossil evidence of their previous distribution. The current population is divided into many separate, small, and geographically isolated subpopulations (Roberts 1988).

Giant pandas face a number of threats. They are completely dependent on bamboo as a food source, which makes them very vulnerable to stochastic events. Bamboo reproduces infrequently, flowering in a mass event (all individuals of the species produce flowers and seeds at the same time) and then the adults die. It may take a number of years before new shoots appear from the seeds. Pandas themselves also reproduce infrequently and have extended parental care, greatly reducing their birth rate. In addition, because China's human population is so large, and the pandas live in areas that have a lot of timber resources, pandas are in danger of losing even more of their habitat as these areas are cleared. All of these factors, and others such as poaching, contribute to the panda's current endangered status.

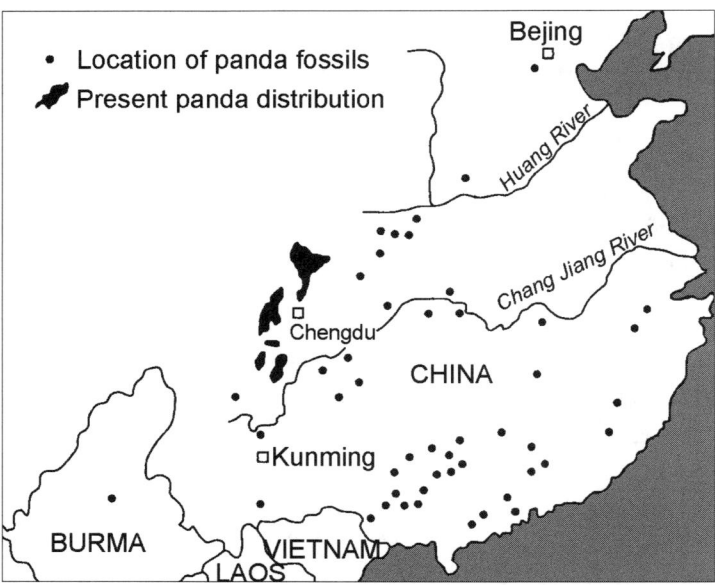

Figure 6.2 Pandas were once found throughout China and even into Burma; they are now restricted to a few areas (modified from Roberts, 1988).

The Chinese government has developed conservation guidelines and established protected areas for the pandas, but the ever-expanding needs of the human population continue to threaten the pandas. Even if the pandas do not lose any additional habitat to logging and agriculture, they are still vulnerable to poaching. Poachers hunt the rare and attractive pandas for their hides, which sell for a lot of money in black markets. The available demographic data from the pandas indicates that their population is growing slightly. However, because so little habitat is available for the pandas, excess individuals have nowhere to go when they mature. In this exercise you will incorporate density dependence and demographic stochasticity to examine the viability of a panda population in the Shuidonggou Valley. You will also look at the affect of even very small amounts of poaching on the population.

Exercise A. Growth in a panda population

Background

In this exercise you will evaluate the population growth and stability of a panda population. You will look at the effects of carrying capacity and demographic stochasticity on the stability of the population over time.

Exercises

1. Open the program RAMAS EcoLab (double-click on its icon with your mouse).
2. Click on *Age and Stage Structure*.
3. Under the *File* menu, select *Open* and choose *GiantPanda.st*. This file contains an age-structured matrix of a giant panda population based on data from Zhou and Pan 1997.
4. Under the *Model* menu select *General Information*. Notice that the *Duration* is set at forty. This simulation will model forty time steps. Each time step in this model is 3 years.
5. Under the *Model* menu select *Initial Abundances*. Notice the structure of the age distribution. In this model, the population is divided into 5 age classes of 3 years each (0–2; 3–5; 6–8; 9–11; 12–14; 14+). Click *Cancel* to exit this window.
6. Now you are going to run the simulation; under the *Simulation* menu select *Run*, or press *Ctrl-R* on your keyboard. At the bottom right corner of this window you will see a message when the simulation is complete. At this point you may close the window by clicking on the X in the upper right corner of the window.
7. Under the *Results* menu, select *Trajectory Summary*. Here you can view the population growth curve. The time scale, like the age classes, is in 3-year time steps. Print the plot of the *Trajectory Summary* by clicking on the printer icon.
8. You can view the numbers corresponding to the plot by clicking the *Show Numbers* button in the upper left corner of the *Trajectory Summary* window. You can return to the plot by clicking the *Show Plot* button also in the upper left corner of the *Trajectory Summary* window. Exit the *Trajectory Summary* by clicking on the X in the upper right corner of the window.

Density Dependence

You are now going to add density dependence to this model. In this population, ecologists believe the remaining habitat is already supporting the maximum number of pandas. Therefore, individuals added to the population will either replace existing individuals that have died, or be forced to leave the area, or die. In the program you will use the contest form of density dependence, meaning some individuals "win" access to a territory and resources, whereas the excess individuals have decreased survival or reproduction.

1. From the *Model* menu select *Density Dependence*. Change the selection for *Density Dependence* type from *Exponential* to *Contest*. Click OK to exit this window.
2. Now you are going to run the simulation without any additional management practices such as removal of additional animals at each time step. Under the *Simulation* menu select *Run*, or press *Ctrl-R* on your keyboard. Close the window by clicking on the X in the upper right corner of the window.
3. Under the *Results* menu, select *Trajectory Summary*. Here you can view the population growth curve. The time scale, like the age classes, is in 3 year time steps.
4. You can view the numbers corresponding to the plot by clicking the *Show Numbers* button in the upper left corner of the *Trajectory Summary* window. You can return to the plot by clicking the *Show Plot* button, also in the upper left corner of the *Trajectory Summary* window.
5. In Table 6.3, record the *Average Abundance* of the population at the final time step. Exit the *Trajectory Summary* by clicking on the X in the upper right corner of the window.
6. Repeat the simulation 10 times and record the values for each simulation in Table 6.3.

Table 6.3. Population abundance of the panda population.

Simulation	Abundance
1	
2	
3	
4	
5	
6	
7	
8	
9	
10	

Questions

1. Describe what the population trajectory looks like before and after density dependence is added.

2. Do we need to be concerned with demographic or environmental stochasticity for this population?

3. Using the data from Table 6.3, how many replications had a final value higher than the initial abundance? How many had a final abundance that was lower? How does density dependence affect the maximum population size?

4. How can you explain this result when the population has a positive growth rate?

5. How will density dependence affect the long-term survival of this population?

Exercise B. The impact of poaching on pandas

Background
Poaching can be another risk facing small populations of endangered species. Poaching of pandas for their skins has become increasingly common despite the large penalties imposed by the Chinese government. Any negative impact on the population will counteract the positive growth rate and push the population into a decline. In this exercise you are going to look at the effect of different intensities of poaching.

Exercises
1. Under the *Model* menu, select *General Information*. Be sure that the box for *Use demographic stochasticity* is selected. Click OK.
2. In the *Model* menu, select *Density Dependence*. Choose *Contest* as the density dependence type to be used. Enter the *Max growth rate* of 1.05 and a *Carrying capacity* of 20. Click OK.
3. Select *Management and Migration* from the *Model* menu. Click on the *Add* button. Select *Harvest/Emigration*. You will see five *Harvest/Emigration* options in the box on the left side of the screen. Each option represents the number of individuals harvested per time step in an age class. The first option is to remove one individual from the 3 to 5 age class. The second option is one individual from the 6 to 8 age class. The third option is one individual from the 9 to 11 age class. The fourth option is one individual from the 12 to 14 age class. The fifth option is to remove one individual from the 14+ age class.
4. Highlight each option and unselect *Ignore this action* from the top of the screen. The *(ignored)* comment next to each option should disappear if you have unselected the *Ignore this action*. The model now will remove 1 individual from each class each time step.
5. From the *Simulation* menu, select *Run*. Close the simulation window when the simulation is finished.
6. Select *Trajectory summary* from the *Results* menu. You can view the numbers corresponding to the plot by clicking the *Show Numbers* button in the upper left corner of the *Trajectory Summary* window.
7. Close the *Trajectory Summary* window and select *Extinction/Decline* from the *Results* menu. You can view the numbers by clicking the *Show Numbers* button as in step 6. In Table 6.4, record the probability of extinction at the end of the simulation. Record the value at the 0 threshold. Rerun the model five times.
8. Highlight the second *Harvest/Emigration* option and then select *Ignore this option*. Highlight the fourth *Harvest/Emigration* option and then select *Ignore this option*. We will now remove one individual from three age classes in each time step.
9. Repeat steps 6 and 7.
10. Highlight the first *Harvest/Emigration* option and then select *Ignore this option*. Highlight the fifth *Harvest/Emigration* option and then select *Ignore this option*. We will now remove one individual from the entire population in each time step as in step 4.
11. Repeat steps 6 and 7.
12. Compute the average time to extinction for each harvesting level by adding together the times to extinction and dividing by 5 (the total number of replications).

Table 6.4. The effect of poaching on population persistence in a panda population.

Poaching 1 per age class per time step	
Simulation	Probability of Extinction
1	
2	
3	
4	
5	
Average	
Poaching 1 from age classes 1, 3, and 5 per time step	
Simulation	Probability of Extinction
1	
2	
3	
4	
5	
Average	
Poaching 1 from entire population per time step	
Simulation	Probability of Extinction
1	
2	
3	
4	
5	
Average	

Questions

1. Does poaching have a noticeable effect on the survivorship of the population?

2. Would you predict that there is any level of harvesting that the population can withstand?

3. Given what you have learned about the giant panda and the threats facing it, what would you recommend to those concerned with its long-term survival?

Your lab report should include the following:

1. Completed Table 6.3 and answers for questions 1 through 5 for the density dependence section of Exercise A
2. Completed Table 6.4 and answers for questions 1 through 3 for the poaching section of Exercise B

References

Gilpin, M. E. and M. E. Soulé. 1986. Minimum viable populations: Processes of species extinction. In M.E. Soulé (ed.), *Conservation Biology: The Science of Scarcity and Diversity*, pp. 19–34. Sinauer Associates, Sunderland, MA.

IUCN Species Survival Commission. 1994. *IUCN Red List Categories*. IUCN Publication Services, Gland, Switzerland.

Reid, W. V. and K. R. Miller. 1989. *Keeping Options Alive: The Scientific Basis for Conserving Biodiversity*. World Resources Institute, Washington, D.C.

Roberts, L. 1988. Conservationists in panda-monium. *Science*, 241: 259–531.

Zhou, Z. H. and W. S. Pan. 1997. Analysis of the viability of a giant panda population *J. Appl. Ecology*, 34: 363–374.

Laboratory 7

Hector's Dolphins and the Red-Cockaded Woodpecker: Conserving Dwindling Populations

Determining the impact of a human activity on a species' population is important for making conservation decisions. The success of different conservation actions may vary, and it is important that we try to predict what actions will be the most successful. In this lab you will assess management options for a population of dolphins off the coast of New Zealand and a population of endangered red-cockaded woodpeckers in Georgia.

Introduction

Money for conservation is limited, so it is often important to predict the effects of conservation actions before we spend time and money on something that might not work. It may also be easier to convince policy makers and granting agencies that a conservation effort is necessary and worthwhile if such predictions are made. Population modeling, as you have seen in the previous labs, is a powerful tool for assessing the status populations and the factors that might affect their long-term survival.

Where Should We Focus Conservation Efforts?

Fecundities, survival rates and numbers of individuals often vary in different age classes. Because of these differences, some age classes may influence population growth more than other age classes. Protecting a given number of individuals in one age class may, therefore, not have the same effect on a population as protecting an equal number in another. Differences in behavior, habitat use, or resource use may occur as an individual develops and grows, allowing conservation strategies to focus on protecting individual age classes. One conservation strategy may be more or less effective than another strategy depending on the distribution of survival and fecundity rates within a population.

A **sensitivity analysis** is often conducted to decide where conservation efforts should be focused. First, a model is run to predict the future population abundance based on current survival and fecundity values. Then, one by one, each value is increased in the model to predict which change will lead to the highest overall increase in future population size. For example, if an increase in the survival rate of 2 year olds leads to the highest final abundance in the model, we might spend more time and money trying to increase the survival rate of 2 year olds in the wild.

You might think that younger classes would be more important for the future of a population because they have a longer life (and a longer reproductive life) than older classes. However, very young individuals often die before they begin reproducing. Alternatively, you might think that the age class with the highest fecundity would be the most important to the future of the population. Often older age classes have higher fecundities, but their reproductive life might not be as long as the younger individuals. It is often difficult to predict which age class is most crucial to the future survival of a population until a sensitivity analysis is actually conducted.

Saving the Loggerhead Sea Turtle

Conservation of the loggerhead sea turtle (Figure 7.1) is a good example of why a sensitivity analysis is important. The loggerhead sea turtle is a threatened marine reptile and many efforts have been made to conserve its populations. The loggerhead spends most of its adult life at sea, coming to land only once a year (or once every few years) to lay eggs. The hatchlings emerge long after the mother has left, using moonlight to guide themselves to their home in the sea.

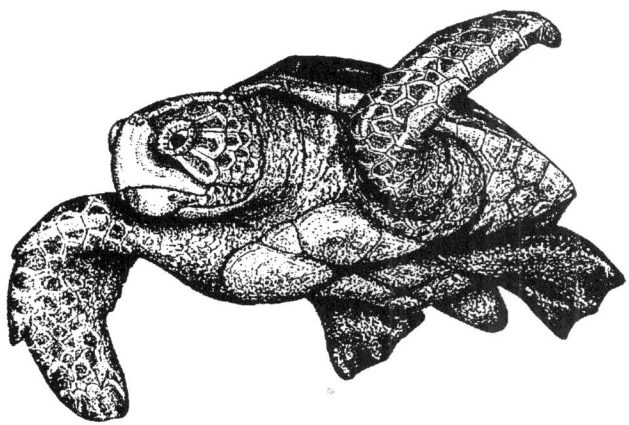

Figure 7.1. The loggerhead sea turtle.

Most conservation efforts for this species have been focused on the eggs, which are quite vulnerable to a variety of predators and to general disturbance. One conservation strategy, designed to increase the survival of the eggs to the hatching stage, is protection of the nesting areas and removal of eggs to protected hatcheries. Although this may be helping the species, it may not be the most efficient way to conserve the sea turtle. Once the eggs hatch, young turtles are very vulnerable to natural predation. Many are eaten before they even reach the water. Although the survival of the young is important, focusing on this age class may not produce the most successful conservation plan.

Increasing adult survival might be more critical for conserving the population. Larger turtles are much more likely to survive to reproduce than eggs or hatchlings. Adults are immune to predation from all except the largest predators, such as sharks at sea and Florida panthers on land. Humans also affect the survival of adult stages. Adults often get captured in shrimp trawls and drown. One way to prevent these accidents is to install escape hatches called "turtle exclusion devices" (known as TEDs) in the fishing gear. These devices can drastically reduce the mortality of larger turtles and may be a more effective way to conserve the population.

Population models can provide insight into which age classes are the most critical for long-term survival. Crowder et al. (1994) measured the survival and fecundity rates of the loggerhead sea turtle. They then used a population model to conduct a sensitivity analysis and found that the long-term survival of the loggerhead population was most sensitive to the survival rate of adult turtles.

Suppose the TEDs were required on all shrimp trawls and this action produced a 10% increase in the adult survival rate. The probability that the population would decline below 20,000 turtles without the TEDs in 30 years is estimated to be approximately 60%. When the TEDs are used, the estimate of the probability of dropping below 20,000 turtles declines to less than 10% in thirty years. Alternatively, suppose that instead of using the TEDs, a management plan that increased hatchling survival 10% was implemented. With the increased hatchling survival, the probability of declining below 20,000 turtles is reduced to 35% in thirty years. In this case, the conservation actions that increase adult survival rather than hatchling survival will have the greater effect on the long-term survival of the population of sea turtles.

Now you will use population modeling to help you evaluate alternative management approaches for two threatened species. In the first exercise you will examine the effects of human activity on an endangered species and simulate population growth before and after a conservation action. In the second exercise you will determine the age class on which conservation efforts should be focused to most cost effectively conserve a dwindling population. In other words, you will conduct a sensitivity analysis.

Exercise A: A wildlife sanctuary for Hector's dolphins

Background

The Hector's dolphin is an endangered species of bottlenose dolphin that is found only in the waters around New Zealand. It has been estimated from aerial surveys that there are about 3000 of these animals remaining. One population that is critically threatened by extinction is located in the Pegasus Bay. In this population there are about 740 animals, and marine biologists believe their numbers are declining rapidly.

Gill nets are commonly used in this area for capturing fish. The dolphins often swim into these nets and become entangled as they struggle to get free. Because dolphins are mammals and need air to breathe, they often drown before they are found. It has been estimated that gill net entanglements cause approximately 57% of dolphin deaths in this population. A proposal has been made to make Pegasus Bay a marine wildlife sanctuary to preserve this endangered species. If this occurs, gill-net fishing will not be allowed in the bay.

Because fishing is an important component of the economy, evidence of this decline and the effects of gill-net fishing will be necessary to convince people that this area should be made a sanctuary for the dolphins. In this exercise you will be asked to determine if this population is really declining and in danger of extinction. You will also determine how gill-net fishing affects the population and whether creating a sanctuary in the bay will allow the population to recover.

Exercises

Part 1: Potential for growth

First you will examine the demographic data on Hector's dolphins to determine the potential growth rates of the population. Using the life table for Hector's dolphin shown in Table 7.1 and RAMAS EcoLab software, we will project the future abundance of the Pegasus Bay population. The survival rates in the table include the effects of gill-net fishing. Assume that any decline below fifty individuals is likely to lead to extinction.

Table 7.1. Life history table for Hector's dolphin generalized from Slooten and Lad (1991).

Age Class	Abundance	Fecundity	Survival Rates
0 - 7 years	214	0	0.36
7 - 14	88	1.51	0.36
14 - 21	36	1.51	0.00

1. Open RAMAS EcoLab (double-click on its icon with your mouse).
2. Click on *Age and Stage Structure*.
3. Under the *Model* menu, select *General Information*. In this window title the file *Hector's Dolphin*.
4. Also in the *General Information* window, change the *Duration* from 0 to 20. By doing this, you have set the program to simulate 20 time steps. Each time step represents 7 years.
5. Also in the *General Information* window, set the number of *Replications* at 1. You will run one simulation to view how the population of dolphins is expected to change over time. Click OK to exit this window.
6. Under the *Model* menu, select *Stages*. Click the *Add* button until you have 3 stages. Under *Name*, enter the age classes found in Table 7.1. Do not worry about the *Average weight* column. Click OK to exit this window.
7. Under the *Model* menu, select *Initial Abundances*. The age classes in the previous step will appear in the window. Under each age class, enter the appropriate number of individuals as listed in Table 7.1.
8. Under the *Model* menu, select *Stage Matrix*. The estimated fecundities and survival rates are listed in Table 7.1. Your matrix should look like Table 7.2. Click OK to exit this window.

Table 7.2. Age matrix for Hector's dolphin.

0	1.51	1.51
0.36	0	0
0	0.36	0

9. Now you are going to run the simulation. Under the *Simulation* menu select *Run*, or press Crtl-R on your keyboard. At the bottom right corner of this window, you will see a message when the simulation is complete. At this point you may close the window by clicking on the X in the upper right corner of the window.
10. Under the *Results* menu select *Trajectory Summary*. Here you can view the population growth curve. The time scale, like each age class, is in 7 year time steps. Before printing your graph, go to *Chart Properties* by clicking on the *scale* button, third from the left, along the top of the window. Give your graph a title and change the x-axis label to say Time Steps (1 step = 7 years). Click OK and then print the plot of the *Trajectory Summary* by clicking on the printer icon.
11. You can view the number of dolphins at each time step by clicking the *Show Numbers* button in the upper left corner of the *Trajectory Summary* window.
12. Fill in Table 7.3 with the abundance at time steps 0, 5, 10, and 20.

Table 7.3. Abundance table for Hector's dolphin.

Time Step	Abundance
0	
5	
10	
20	

Questions

1. How many years does each time step represent in this population model?

2. What does this model predict about the future of the Hector's dolphin population if nothing is done?

Part 2: The elimination of fishing pressure

As mentioned earlier, gill-net fishing causes about 57% of the mortality experienced by the dolphins in this area. The remaining 43% of deaths are due to other causes, such as predation (by species other than humans), sickness, and old age. In this section, you will be asked to determine whether the dolphin population is likely to recover if Pegasus Bay is made a wildlife sanctuary. Assuming that fishing entanglements represent the major mortality source, if the population still declines after this source of mortality is removed, it is likely that some other factor is also limiting the growth of the dolphin population.

1. You will be using the same file you created to simulate the Hector's dolphin population. You will adjust the mortality levels to predict the consequences of eliminating entanglement mortality. Under the *Model* menu select *Stage Matrix*.
2. Although we believe that entanglements cause 57% of the mortality in this population, we will be conservative and remove only 50% of the mortality (this will make our argument stronger if someone argues that 57% is an overestimate of the mortality caused by entanglements). To calculate the new mortality rate, subtract the survival rate from 1.0 and multiply this new number by 0.5. A new survival rate can them be calculated by subtracting the mortality rate from 1.0. Make this calculation for both survival rates in the table.
3. In the *Stage Matrix* window of the *Model* menu, update your stage matrix with the new values and click OK.

4. Under the *Simulation* menu select *Run* or press *Ctrl-R* on your keyboard. Close the simulation window when the simulation is complete.
5. View your *Trajectory Summary* from the *Results* menu, as before, and print a copy by pressing the *Print* button, third from the right on the tool bar.
6. Select the *Show Numbers* button to view abundances. Fill in Table 7.4 with the average abundance at time steps 0, 5, 10, and 20.

Table 7.4. Abundance table with 50% mortality removed.

Time Step	Average Abundance
0	
5	
10	
20	

Questions

1. How does the population trajectory change when fishing entanglements are removed?

2. Estimate how many times larger the population will be at time steps 5, 10 and 20 as the fishing mortality is eliminated. Do this by dividing the abundance without fishing by the abundance with fishing (this is called a ratio). Fill in the values in Table 7.5.

Table 7.5. Comparison of abundance with and without fishing pressure.

Time Step	Abundance without Fishing	Abundance with Fishing	Ratio
5			
10			
20			

3. Based on your results, would you recommend that Pegasus Bay be designated a sanctuary where no gill-net fishing is allowed? Why or why not?

4. If fishing mortality is removed, as you simulated in this model, do you think the population would increase over the 20 time steps as much as it did in this model? (Hint: Think about the lessons you learned in Laboratory 3.)

Exercise B: Conservation of the red-cockaded woodpecker

Background

Determining what is wrong with a population and how to reverse a population decline is not always as simple as the dolphin situation. Sometimes conservation biologists need to determine the most efficient way to help a population. A population may respond better if we focus conservation efforts on a particular age or stage class.

The red-cockaded woodpecker is an endangered bird that has very specific habitat requirements. This species builds nest holes in older long-leaf pine trees throughout the southeastern United States. Unfortunately, this is also a favored location for lumber extraction. Heavily logged areas tend to lack the older trees that the woodpecker uses. In some heavily managed forests, artificial nest boxes are being used to replace the lost nesting sites.

In this exercise you will use an age-based population model for a red-cockaded woodpecker population in North Carolina. Data for this model was modified from Reed et al. 1988. To examine the contribution of each age class you will alter survival rates of different age classes and estimate the effects these changes would have on the long-term abundance of the population. In other words, you will conduct a sensitivity analysis.

Exercises

Part 1: Baseline model

The baseline model uses the present estimated survival and fecundity rates in a population. We use the term "baseline" because we are going to use the results of this model to establish a basis for comparison to evaluate different conservation strategies.

1. Open the program *RAMAS EcoLab* (double-click on its icon with your mouse).
2. Click on *Age and Stage structure*.
3. Under the *Model* menu, select *Open* and choose *Woodpecker.st*. Note that this model includes the effects of demographic stochasticity and runs 100 replications.
4. Open the *File* menu and select *Stage matrix*. Familiarize yourself with the values there. Note that each age class represents 1 year. In Table 7.6, fill in the survival values for each age class for the baseline model. Click OK to exit this window.
5. Under the *Simulation* menu select *Run*, or press *Ctrl-R* on your keyboard. You can speed up the simulation by clicking the *Display Simulation Text* button in the window (top left). A window will pop up telling you that the simulation is complete. Click on the X in the upper right corner to close the window.
6. Select *Trajectory Summary* from the *Results* menu. This will show a graph of the population size versus time. Print this graph by pressing the *Printer* icon.

7. View a table of numbers that gives you the population size at each time step. Click on the *Show Numbers* button, in the *Trajectory Summary* window. Take note of the final average abundance.

Questions

1. Describe the population trajectory graph for the baseline model.

2. How many years does each time step in this model represent?

3. What is the average population size after 20 years? What is it after 50 years?

Part 2: A sensitivity analysis

Now you will increase the survival rates of each age class by 5%, 1 age class at a time, to see which change will have the greatest effect on the population.

1. Calculate a 5% increase in survival rates over the baseline values (multiply by 1.05) and fill in the Table 7.6 for each stage.

Table 7.6. Survival rate table for red-cockaded woodpecker.

Age Class	Survival Rate	
	Baseline	+5%
0		
1		
2		
3		
4+		

2. Select *Stage Matrix* from the *Model* menu. To change a number in the matrix, double-click on the number and enter the new value. Increase the survival rate of the 0 year olds (in the second row, first column) by 5%. Click OK.
3. Rerun the simulation with the new survival rate. In Table 7.7, write the average population abundance after 20 years.
4. Go back to the *Stage Matrix* and change the survival rate of the 0 year olds back to the original baseline value. Then change the survival rate of 1 year olds to reflect a 5% increase, using the value from Table 7.6. You may save this file if you wish, but make sure to use a different name, because you will need the original file for the next exercise.
5. Rerun the simulation with the new survival rate. Write the average population abundance after 20 years in Table 7.7.
6. Repeat steps 4 and 5 with, in turn, a 5% increase in the survival rate for 2 year olds (fourth row, third column), the 3 year olds, and the 4+ individuals. Make sure you reset all of the survival values back to the baseline values, except the one you are increasing.
7. Fill out Table 7.7 listing the average population size after 20 years under each of the new survival rates you have entered.

Table 7.7. The survival rate and average population size after 20 years for each stage after changes to the model.

Stage	Change	Survival Rate	Average Population Size after 20 Years
0	+5%		
1	+5%		
2	+5%		
3	+5%		
4$^+$	+5%		

Questions

1. What is the effect of increasing each survival rate by 5%? Is it the same for all age classes? Which increase in survival led to the largest average population size after 20 years?

2. Based on your results, what management actions would you recommend for conservation of this population?

Part 3: An alternate view

It is often easier to produce a change in the youngest age class of a species than in any other age class. This is because the youngest individuals do not move around as much as older individuals and can easily be monitored or taken into captivity. In this example, the young woodpeckers could easily be monitored in their nests or even taken into captivity and given special attention. Therefore, the amount of effort (and money) expended in increasing the survival rate of 0 year olds by 25% might be comparable to the amount of effort expended in increasing the survival rate of older age classes by only 10%. In this exercise you will increase the survival rate of 0 year olds by 25% and increase that of the older age classes by 10% and compare the results.

1. Calculate a 25% increase over the baseline value in the survival rate of 0 year olds (multiply by 1.25) and enter the value in Table 7.8.
2. Calculate a 10% increase over the baseline values of all older age classes (multiply each by 1.10) and enter the new values in Table 7.8.
3. Reopen the file *Woodpecker.st*. Make sure it has the original baseline values of all the survival rates.
4. Change the survival rate of 0 year olds to the new value you've calculated, and rerun the simulation. Record the average population size after 20 years in the Table 7.8.
5. Reset the survival rate of 0 year olds to the original baseline value, and increase the survival rate of 1 year olds to the new value in Table 7.8. Rerun the simulation and record the average population size after 20 years in Table 7.8.
6. Repeat this process, changing the survival rate of each age class by the correct amount, making sure to reset all other survival rates to the baseline values, and record the average population size after 20 years in Table 7.8.

Table 7.8. The survival rate and average population size after 20 years for each stage after changes to the model.

Stage	Change	Survival Rate	Average Population Size after 20 Years
0	+25%		
1	+10%		
2	+10%		
3	+10%		
4^+	+10%		

Questions

1. Which increase in survival led to the largest average population size after 20 years?

2. Based on these results, how would you change your answer to question 2 from Part 2, in which you added the same percentage to each survival rate? Discuss this in terms of the amount of conservation effort expended in changing each survival rate.

Your lab report should include the following:

1. One population trajectory graph, 1 age matrix (Table 7.2), 1 abundance table (Table 7.3), and answers to questions 1 and 2 for Part 1 of Exercise A
2. One population trajectory graph, 1 abundance table (Table 7.4), and answers to questions 1 through 4 for Part 2 of Exercise A
3. One population trajectory graph and answers to questions 1 through 3 for Part 1 of Exercise B
4. One survival rates table (Table 7.6), 1 abundance table (Table 7.7), and answers to questions 1 and 2 for Part 2 of Exercise B
5. One abundance table (Table 7.8) and answers to questions 1 and 2 for Part 3 of Exercise B

References

Crowder, L.B., D. T. Crouse, S. S. Heppell, and T. H. Martin. 1994. Predicting the impact of turtle excluder devices on loggerhead sea turtle populations. *Ecological Applications*, 4: 437–445.

Reed, J.M., P. D. Doerr, and J. R. Walters. 1988. Minimum viable populations size of the red-cockaded woodpecker. *Journal of Wildlife Management*, 52: 385–391.

Slooten, E., F. Lad. 1991. Population biology and conservation of Hector's dolphins. *Canadian Journal of Zoology*, 69 (6): 1701–1707.

Laboratory 8
African Market Hunting and Tuna Exploitation: Maintaining Sustainable Levels of Harvesting

Overexploitation is one of the most pervasive issues in conservation biology. Poverty and profit drive people to remove more individuals from populations than can be replaced naturally. Stopping this trend is often more complicated than passing laws. You will examine some of the implications of this behavior for a number of populations.

Introduction

Throughout history, humans have harvested natural resources for food, shelter, and commerce. For centuries, this practice of harvesting had only minor effects on the species being harvested because populations were able to replace themselves faster than humans could remove them. The growth of the human population and the improvement of technology have led to unsustainable harvesting of many species. When individuals are removed from a population faster than the population can replace them through reproduction, the population begins to decline. If the rate of exploitation is not reduced, the harvested population will eventually become extinct. Sustainable harvesting practices require careful management so that species being used as resources continue to thrive and are available for future generations.

The Goal of Sustainable Harvesting

To achieve the goal of sustainable harvesting, a balance is needed between over- and underexploitation of natural populations. **Overexploitation** is the removal of so many individuals that the population faces danger of extinction. The detrimental effects are felt both by the exploited population and by the harvester who may depend on this population for economic reasons. **Underexploitation** is the removal of fewer individuals than a population can withstand. How can harvesters, maximize their yield without threatening the viability of a harvested population?

As we have already discussed, the typical trajectory of population growth under density dependence has several phases. During the first phase, when population numbers are low, the population size increases slowly. During the phase of population expansion, the number of individuals added at each time period continues to increase. At some point, however, density-dependent factors reduce the growth rate of the population and the total number of individuals begins to level off. Recall that the shape of the growth curve may differ under various types of density dependence (scramble, contest, or ceiling). However, the highest rate of growth is always at intermediate levels of population density.

At low densities populations increase slowly, and it takes a long time to recover from a harvest. At high densities populations suffer from the effects of intraspecific competition and respond slowly to a slight reduction in numbers. At intermediate densities, however, the population is experiencing its highest growth rate and will be able to replace harvested individuals quite rapidly.

Based on this pattern of response, biologists have proposed a concept called **maximum sustainable yield** (MSY), which is the largest number of individuals that can be removed from a population over time without causing population decline. This idea is based on the assumption that sustainable yield can be maintained by harvesting at or slightly under the rate of population growth. Therefore, the maximum sustainable yield can be attained when harvesting rates are maintained close to the highest maximum growth rate that a population can experience (given the carrying capacity of the area it inhabits).

In practice, maximum sustainable yield is often difficult to determine and even more difficult to achieve. Reasons for this include errors when estimating population parameters, unpredictable environmental variation (environmental stochasticity), poor enforcement of harvesting laws, and changes in the number of individuals in age and/or sex classes (population structure) due to harvesting procedures. Often, after estimating the maximum sustainable yield, wildlife managers recommend a slightly lower harvest rate to allow for unpredictable variation.

Different populations will have different maximum sustainable yields. Populations in which individuals mature quickly can usually withstand high harvesting levels. Many of the agricultural crops fit this pattern. Wheat, corn, and rice, to name a few, are species in which individuals live for only 1 year and then die. We can, therefore, harvest all the individuals at the end of a growth season, as long as we leave enough seeds for new individuals to replace them the following spring. Species that produce offspring in large numbers are also less sensitive to high harvest levels. The tenrec of western Madagascar is a small insectivorous mammal, similar to a hedgehog that produces up to 31 young per litter (the largest litter reported for a mammal). Many people of western Madagascar rely on hunting to provide protein in their diet. The tenrec is hunted extensively during part of the year but does not seem to be negatively affected because it has such a high fecundity rate (Ganzhorn et al. 1996).

When individuals take a long time to mature, or when each individual can only produce a few young, harvesting a large fraction of the population will result in large population declines. Large mammals, many birds, large fish, and trees are all groups of organisms with life histories that are not compatible with heavy harvesting, because they have long maturation periods and low replacement rates.

Harvesting Practices

A number of different strategies for harvesting plants and animals have been used by human civilizations. One extreme is to harvest as many individuals as demand dictates, without regard for the biology of the species. When a species is abundant and demand is low, this will have little impact on the population. Unfortunately there have been several cases in which demand exceeded the recovery rate of a harvested population, and the species was driven to extinction.

At the end of the nineteenth century, for example, hunting of the dodo bird was completely unregulated. Sailors and traders heavily harvested these birds when they stopped at the island of Mauritius where the dodo lived. Dodo birds faced an additional threat of introduced nest predators (pigs and monkeys) that were brought along by sailors and traders. Eventually, the entire population of dodos was depleted. On Mauritius and nearby islands, several species of giant tortoise were also driven to extinction because of unregulated overexploitation.

Fixed quotas

Some harvests are based on the collection of a fixed quota of individuals. Using this type of harvest, a set number of individuals is removed from the population at each time period. This strategy is used by several management agencies, including some fisheries. On a specified day of each year, the fishery is opened and the total accumulated catch is recorded on a daily basis. When the quota is reached, the fishery is closed for the rest of the year.

Even with fixed quotas the persistence of a harvested population is not guaranteed. In the 1960s the Peruvian anchovy fishery was one of the largest fisheries in the world, comprising 18 percent of the world's total harvest of fish. The MSY for this population was around 10 million tons annually. For a period of 8 years the catch was close to the MSY, but in 1972 the population crashed. The rapid decline in the population may have been due to a combination of factors, including overfishing by increasingly efficient fleets and the severe environmental phenomenon known as El Niño. Following the decline, the anchovy population might have been saved if fishing had been limited. However, the Peruvian government allowed fishing to continue at the fixed quota because of the strong dependence of the country on the anchovy industry (Figure 8.1).

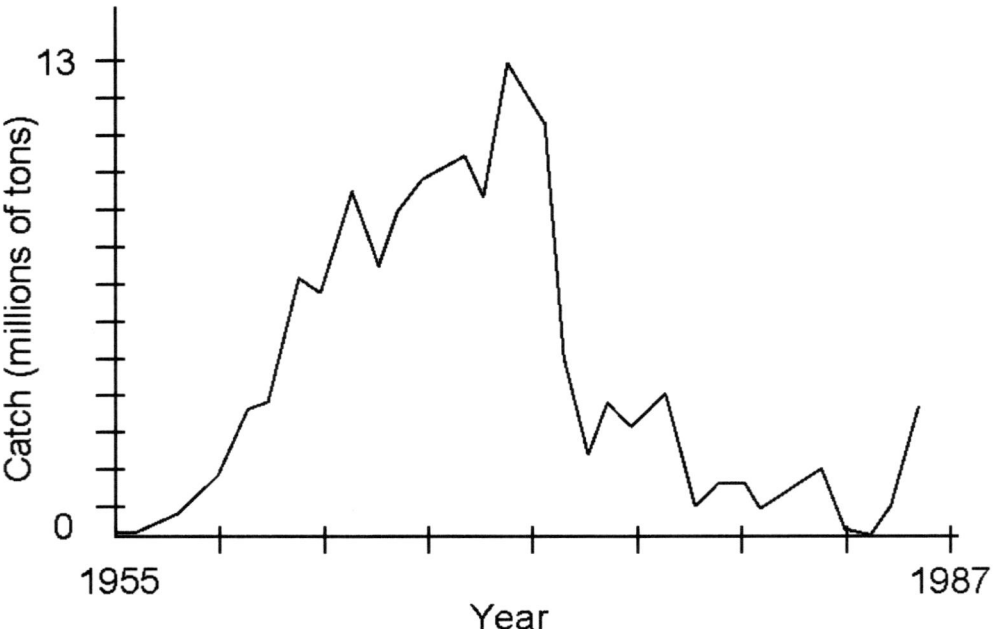

Figure 8.1. Pattern of annual yield of anchovies by the Peruvian anchovy fishery. In 1972 the population crashed, yet a fixed quota continued to be used and harvesting continued (Hilborn and Walters, 1992).

Alternatives to fixed quotas

Some harvesting practices adjust for fluctuations in a population's size by changing the harvesting effort seasonally or annually. From a purely economic standpoint, if a season has a lower yield than predicted, the response is to increase the harvesting effort the following year to make up for the lack. However, from a sustainable harvesting point of view, if normal harvesting effort results in a low yield, a population must have dropped below its optimal MSY. A decrease in harvesting effort the following year will allow the population to recover and favor the future health and persistence of the population.

One specific alternative to a fixed quota is to harvest a fixed percentage of a population each time period. With this strategy, when a population falls to a low abundance, fewer individuals are harvested than when the abundance is high, which can allow the population to recover to higher numbers. By adjusting harvesting effort from season to season, many natural populations have been managed quite successfully. Many fish and game services annually regulate the number of hunting permits issued for deer, elk, and bear. This system is not foolproof, however. For some fish species, such as sardines and herring, individuals form schools when their density is low. Even when a population size is critically low, a single catch may bring in a "normal" yield because a dense school was collected. This can cause severe damage to an already threatened population. As in most cases, knowledge of the biology of a species is the best defense against mismanagement.

Another alternative to fixed quotas is to harvest only those individuals that are of a particular sex or in a certain life stage. Recall from Laboratory 7 that a population is more sensitive to changes in the vital rates of certain age/stage classes than others. By prohibiting the capture of individuals that are important in producing future generations, a wildlife manager can help to ensure the long-term success of the population. This is why hunting permits often are issued for hunting only males of a certain age and why people who engage in recreational fishing can take only fish above some minimum size.

Poaching: A Threat to Sustainable Harvesting

Illegal harvesting often complicates prevention of a decline in a managed species. Although national and international regulations prohibit harvesting more than a set number of animals, or harvesting animals that are considered endangered, it can be difficult to prevent poaching (illegal harvesting) if there is a strong demand for products from a species. A well-known example of illegal trade in products from a managed population is the elephant in Africa. As populations of this species began to crash, it became obvious that international demand on elephant ivory spurred poaching. An international ban on ivory trade helped to reduce some of the incentive for poachers. Now, with elephant populations recovering in some areas, there is pressure to lift the ban on ivory. However, wildlife managers are well-advised to be conservative when issuing permits to account for the inevitable return of poaching.

Calculating Maximum Sustainable Yield

The first step in calculating the MSY of a species is to estimate the maximum number of offspring, or recruits, that can be produced by that species per time step. Recall that this is expected to occur at the point of maximum recruitment in a population. Often, determining the point of maximum growth is simplified by assuming that maximum recruitment occurs at 50% of the carrying capacity (it might not, depending on the growth curve of the population). This rate of recruitment is also the MSY, if we assume that recruited individuals replace those that are harvested. This assumption has some problems that we will discuss later. The following formula is used to determine the MSY for each population:

$$\text{MSY} = \left(\frac{K}{2} \times R\right) - \frac{K}{2}$$

Where K = carrying capacity and R = growth rate

First we calculate the number of individuals that we are trying to maintain in the population from one time step to the next. As we decided above, this number is 50% of the carrying capacity ($K \div 2$). Next, multiply this number by the growth rate (R), to determine how many individuals will be in the population in the next time step. Then, to calculate MSY, simply subtract the number of individuals that were present initially ($K \div 2$), from the population size in the next generation ($[K \div 2] \times R$). This gives you the number of individuals in excess of the number we are trying to maintain after one time step.

If harvesting occurs at the same rate at which a population produces new offspring, we expect that a population will remain stable. However, this is based on the assumption that each individual contributes equally to population growth. As you learned in Laboratory 7, this rarely the case. Harvesting under this assumption combined with the affects of stochastic events (demographic or environmental), may result in a decline of the population. The longevity, or average life span, of a species will also affect the impact of harvesting on a population. Short-lived species, for example, may be less sensitive to harvesting because the animals that are harvested are likely to die from natural causes anyway, in contrast to long-lived species, which are usually quite sensitive to harvesting.

Because of these problems, ecologists Robinson and Redford (1991) suggested that harvesting should only take a percentage of the calculated maximum sustainable yield and that this percentage should vary according to the longevity of the species. They assumed that harvesting could safely take 60% of the calculated MSY of very short-lived species (<5 years), 40% of the MSY of short-lived species (5 to 10 years), and 20% of the MSY of longer-lived species (>10 years). These new values will be referred to as the "adjusted maximum sustainable yield."

In the following exercises, you will look at two different examples of exploitation. In the first, you will examine the rates at which several mammal species are hunted in the forests of Equatorial Guinea. In the second exercise you will investigate the population decline of the western blue fin tuna. In both cases, the harvest rates on species appear to far exceed the maximum sustainable yield. You will be asked to consider techniques for recovery of these species, keeping in mind the needs and practices of the people who depend on them.

Exercise A: Harvesting wild game in Africa

Background

In many tropical countries of Africa, people use wild game as a source of protein. Overexploitation of a species will lead to population decline and eventual extinction of that species. It will also stress the population of humans who depend on these resources for sustenance. Information from the markets where the meat is sold is useful in determining whether any of the species are being overharvested.

In this exercise you will determine whether different mammal species are being hunted sustainably on Bioko Island in Equatorial Guinea. Island populations are often very sensitive to overexploitation because of their limited size and isolation. As you will learn in a future lab, isolated areas have less immigration and a lower carrying capacity, making populations more sensitive to harvesting than populations from larger mainland areas.

The next step will be to use the maximum sustainable yield of each species you calculated to determine how many individuals might be harvested without hurting the population. In this way you will see which populations are being overharvested and which populations could withstand more harvesting pressure. People can then use this information to harvest these resources sustainably, without depleting the populations. You will use the assumptions of Robinson and Redford (1991) to determine the adjusted maximum sustainable yield of each of the mammal species found in the markets of Equatorial Guinea.

Exercises

Part 1: Maximum sustainable yield

1. Table 8.1 is a chart of market hunted mammals from Bioko Island (data from Fa et al. 1995). Calculate the maximum sustainable yield for each species using the information in the columns titled K÷2 and R.
2. Write the values you calculated in the column labeled MSY.

Table 8.1. Commonly harvested wild game species of Africa.

Species	K÷2	R	MSY	Longevity (years)	Adjusted MSY	Observed Harvest	Proportion of Adjusted MSY Taken
Primates							
Patas monkey	24,910	1.17		30.8		781	
Gentle monkey	22,983	1.11		30.8		254	
Crowned monkey	14,524	1.01		28.0		52	
Pruess's monkey	9,984	1.11		30.8		196	
Black colobus monkey	20,573	1.15		30.5		514	
Drill	6,758	1.20		28.6		551	
Red colobus monkey	157,628	1.15		30.0		348	
Pholidota							
Tree pangolin	10,992	2.01		13.1		192	
Rodents							
Bush tailed porcupine	55,468	1.82		22.9		1,581	
Giant pouched rat	135,139	2.01		7.8		2,419	
Artiodactyla							
Ogilby's duiker	13,110	1.26		8.0		3,181	
Blue duiker	22,792	1.63		7.0		732	

Questions

1. A population growth rate of 1.20 is likely to produce a population that is neither growing nor declining with normal levels of environmental and demographic stochasticity. A population with a growth rate of less than 1.20 and that is subject to stochasticity is likely to decline. How many of the animals listed Table 8.1 of hunted mammals have growth rates greater than or equal 1.20?

2. In general do those populations with a larger growth rate have higher maximum sustainable yield values in the table?

Part 2: Adjusted maximum sustainable yield

1. The longevity of each of the species is also given in Table 8.1. Calculate the adjusted maximum sustainable yield of each species by multiplying the maximum sustainable yield values by 0.6 (species <5 years), 0.4 (species 5 to 10 years), or 0.2 (species >10 years). Write the harvest values in the column labeled *Adjusted MSY* in Table 8.1.
2. Next find the proportion of the adjusted maximum sustainable yield harvested by the people by dividing the observed harvest by the adjusted maximum sustainable yield. Write these values in the column labeled *Proportion of adjusted MSY taken*. If this number is greater than 1, the population is being overharvested.
3. Make a bar graph of the proportion of adjusted MSY of each species taken by hunters on Bioko Island.

Questions

1. Why does the maximum sustainable yield of a population occur at 50% carrying capacity and not at 100%?

2. Why might the people of Bioko be interested in sustainable harvesting?

3. Are any of the species on Bioko Island being overharvested? If so, which ones, and what proportion of the adjusted maximum sustainable yield is being harvested?

4. The people of Bioko Island depend on these resources for protein. If they decrease their harvesting rates on some species to prevent their populations from crashing, they must somehow compensate for this loss. How can they change their hunting practices to meet this need? (Hint: Use information from Table 8.1.)

Exercise B: Harvesting bluefin tuna

Background

The bluefin tuna (*Thunnus thynnus*) is an endangered species. Individual bluefin tunas can weigh up to 1500 lb (680.4 kg) and can be sold for up to $30,000 in Japanese fish markets. Because of its high market value, there are still overly high quotas placed on fishing bluefin tuna, even though this species has been extensively studied and monitored. Over the past 20 years, the stock of western Atlantic bluefin tuna has been depleted by over 90%. Fishing intensity has increased to try to maintain catches. Even with the increased efforts, however, catches are substantially lower than in past years: both the tuna and the people harvesting them are losing out in this situation.

In the 1960s a commission called the International Commission for the Conservation of Atlantic Tuna (ICCAT) was formed to protect and monitor the fishing stocks. However, market pressure from the fishing nations of the United States, Canada, and Japan has kept fishing regulations at a minimum. The International Union for the Conservation of Nature (IUCN) has been pressured by the tuna fishing nations not to designate the western Atlantic population as an endangered species despite recommendations by biologists and nonfishing nations to place it on the endangered species list.

In the following exercises we will look at the population dynamics of the western Atlantic bluefin tuna to understand the dynamics of an overharvested species. The first exercise reflects the tuna population found in the western Atlantic nearly 20 years ago. The values for initial abundances, fecundities, and survivals, as well as the intensity of bluefin tuna harvest, are all based on historical figures (obtained from ICCAT 1997). Next you will investigate the effect that a fishing ban would have had on the population. Finally, you will examine the tuna population of today and determine the maximum sustainable yield for this much reduced population.

Exercises

Part 1: The history of the bluefin tuna

In this first exercise you will look at the population growth curve of the western Atlantic bluefin tuna using population estimates from the 1970s. You will observe how different harvesting practices affect the population growth trajectory.

1. Open the program *RAMAS* EcoLab (double-click on its icon with your mouse).
2. Click on *Age and Stage Structure* icon.
3. Under the *File* menu, select *Open* and choose *Bluefin1.st*. This file contains an age-structured matrix based on the 1970 population census. In this model each age class represents 2 years.
4. Under the *Model* menu select *General Information*. Notice that the *Duration* is set at 40 time steps (or 80 years). Also notice in the *General Information* that the model is only set to run 1 replication. More than 1 replication is not necessary because we are not using demographic stochasticity (make sure the box for use demographic stochasticity is **not** checked). Click OK to exit this window.
5. Under the *Model* menu select *Initial Abundances*. Notice the structure of the age distribution. Adults in the 10+ age class have lower abundances than the other age classes. These are the individuals that are highly sought by fishermen.
6. Under the *Model* menu select *Stage Matrix*. Inspect the matrix. Note that all of the age classes have high survivorship without fishing mortality.
7. From the *Model* menu select *Management and Migration*. The value listed here is approximately the level at which bluefin have been harvested since 1970 (notice that it is a proportional harvest, so a constant percentage is taken from the selected age classes, age 0-1 through age class 10+). Make sure there is **not** a checkmark in the box next to *Ignore this action*. Click OK to exit this menu.
8. Now you are going to run the simulation. Under the *Simulation* menu, select *Run* or press *Ctrl-R* on your keyboard. At the bottom right corner of this window you will see a message when the simulation is complete. At this point you may close the window by clicking on the X in the upper right corner of the window.
9. Under the *Results* menu select *Trajectory Summary*. Here you can view the population growth curve. The time scale, like each age class, is in 2 year time steps. Print the plot of the *Trajectory Summary* by clicking on the printer icon.

10. You can view the numbers corresponding to the plot by clicking the *Show Numbers* button ▥ in the upper left corner of the *Trajectory Summary* window. Record the abundances in Table 8.2. Exit the *Trajectory Summary* by clicking on the X in the upper right corner of the window.

11. Now we will investigate what the growth trajectory for the tuna population would have been if there were no fishing from 1970 until the present. From the *Model* menu select *Management & Migration*. Click in the box next to *Ignore this action*. Make sure the box now contains a checkmark. You are now going to run the model with no harvesting from the population.

12. Now you are going to run the simulation. Under the *Simulation* menu select *Run* or press *Ctrl-R* on your keyboard. To close this window, click the X in the upper right corner when the simulation is complete.

13. Under the *Results* menu, select *Trajectory Summary*. Here you can view the population growth curve. The time scale, like the age classes is in 2 year time steps. Print out the trajectory summary by clicking on the printer icon.

14. View the numbers corresponding to the graph by clicking on the *Show Numbers* button in the upper left corner of the *Trajectory Summary* window. Fill in Table 8.2 with the abundance estimates for each of the time steps.

Table 8.2. Abundance of bluefin tuna with and without harvest.

	Abundance	
Time Step	With Harvest	Without Harvest
1		
10		
20		
30		
40		

Part 2: Sustainable harvest of the bluefin tuna

In the first exercise you looked at the abundance of tuna with and without harvesting. Now you will determine the amount of tuna harvested in each time step since 1970. RAMAS EcoLab allows us to estimate the harvest weight of tuna under the harvesting regime that has been used by the tuna fisheries. You will then compare the actual harvest to the harvest we would expect if you had been using sustainable practices.

1. You first want to incorporate harvesting into the model again. Under the *Model* menu select *Management & Migration*. Make sure there is **not** a checkmark in the box next to *Ignore this action*. Click OK to exit this window.

2. You will rerun the simulation under the historical harvesting practices to record harvest weights at each time step. From the *Simulation* menu select *Run* or press *Ctrl-R* on your keyboard. Close the simulation window by clicking the X in the upper right corner of the window.

3. Under the *Results* menu select *Harvest Summary*. To view the numbers corresponding to the graph, select the *View Numbers* icon in the upper left of the *Harvest Summary* window. Complete Table 8.3 with the number of kilograms of fish harvested at each time step. Also include the total kilograms of fish harvested found on the last line of the *Harvest Summary* table.
4. Now, through trial and error, you will determine the percentage of harvest from age classes 0-1 through 10+ which represents the maximum sustainable harvest the western bluefin tuna can sustain. Under the *Model* menu, select *Management & Migration*. Enter a new proportion of individuals you would like to remove from the population in each step. Click OK to close this window.
5. Now you will run the simulation to see if the new harvest proportion is sustainable. Under the *Simulation* menu, select *Run* or press *Ctrl-R* on your keyboard. Close the simulation window by clicking on the X in the upper right corner.
6. Under the *Results* menu, select *Trajectory Summary*. If the population is growing each time step, the harvest proportion is too low. In the *Management & Migration* menu increase the proportion of individuals harvested. Rerun the simulation. If the population is declining, the proportion of individuals harvested is still too high. Reduce the harvest proportion and rerun the simulation.
7. Repeat steps 4 through 6 until the population abundance appears to be relatively constant. This harvest proportion is the maximum sustainable harvest. Under the *Results* menu, select *Harvest Summary*. Complete Table 8.3 with the weight of tuna harvested at each time step under this sustainable regime. Also include the total kilograms of tuna harvested.

Table 8.3. Harvest values for 2 different harvesting strategies.

	Amount of Tuna Harvested (Kg)	
Time Step	Actual Harvest	Sustainable Harvest
1		
10		
20		
30		
40		
Total (kg)		

Questions

1. Is the population growing or declining when no harvest is included in the model? From the trajectory graph you printed, would you predict that tunas could maintain their population size when being heavily fished?

2. Is the population growing or declining under the model with 1984 harvest rates? Is this level of harvest sustainable?

3. How long (in years and time steps) did it take the population to drop to half of the original population under the actual harvesting levels? How long did it take to decline to one-tenth of the original population?

4. What is your estimate of the maximum sustainable yield (in terms of proportion harvested at each time step from the selected age classes)?

5. In the first few time steps of sustainable harvesting, is the harvest larger or smaller than with the actual rate of harvest (in 1984)?

6. When does the yield of the actual rate of harvest drop below the yield of sustainable harvest?

7. In the long run (over 40 time steps), which level of harvesting would have produced a greater yield? What would you tell tuna harvesters to convince them to change their harvesting practices?

Part 3: Management options for the future

In this exercise you will explore different options for managing the tuna fisheries using current abundance estimates. You will use 3 different harvesting levels to compare how population abundance and harvesting yields change over time. This information can be used to support different harvesting regimes.

1. Open the program *RAMAS EcoLab* (double-click on its icon with your mouse).
2. Click on *Age and Stage Structure.*
3. Under the *File* menu, select *Open* and choose *Bluefin2.st*. This file contains an age structured matrix based on the 1994 population census. In this model each age class represents 2 years. Compare the original abundance to those in the previous exercise.
4. For the first simulation, you will look at the future population growth of the Western bluefin tuna if fishing was banned. Under the *Model* menu, select *Management & Migration*. Make sure there is a checkmark next to *Ignore this action*. Click OK to close the window.
5. You will now run the simulation. Under the *Simulation* menu select *Run*, or press *Ctrl-R* on your keyboard. When the simulation is complete close the window by clicking the X in the upper right corner.
6. From the *Results* menu, select *Trajectory Summary*. Print the trajectory summary graph by clicking the printer icon.
7. To view the numbers corresponding to the graph, select the *Show Numbers* icon in the upper right corner of the window. Complete Table 8.4 with the population abundance at each time step.
8. The bluefin tuna fishery is very profitable. It is, therefore, unlikely that you will eliminate all fishing from the bluefin population. You will now add various levels of harvesting to see how they affect the future population growth rate and harvest levels. From the *Model* menu, select *Management & Migration*. Remove the checkmark from the box next to *Ignore this action*. Note that the harvest level is very low (make sure the *Proportion of Individuals* reads 0.01); you will remove only 1% of the population at each time step. Click OK to close this window.
9. You will now run the simulation; from the *Simulation* menu, select *Run*. When the simulation is complete close the window.
10. Under the *Results* menu, select *Trajectory Summary*. Print the trajectory graph.
11. View the numbers associated with the graph by clicking the *Show Numbers* icon in the upper left corner of the window. Record the population abundance for 1% harvesting in Table 8.4.

Table 8.4. Abundances without harvest and with 2 different rates of harvest.

	Abundance		
Time Step	Without Fishing	With 1% Harvest	With 3.9% Harvest
1			
10			
20			
30			
40			

12. Under the <u>R</u>esults menu select <u>H</u>arvest Summary. Click on the *Show Numbers* icon to view the weights of tuna harvested. Record the harvest amounts for each time step in Table 8.5.
13. You will now increase the harvesting level to 3.9%. Under the <u>M</u>odel menu select *Management & Migration*. Change the *Proportion of individuals* in the lower left corner to 0.039. Click OK to exit this window.
14. Now you will run the simulation; from the <u>S</u>imulation menu select <u>R</u>un. When the simulation is complete, click the X in the upper right corner to close the window.
15. Under the <u>R</u>esults menu, select <u>T</u>rajectory Summary. Print the trajectory graph by clicking on the printer icon.
16. View the numbers corresponding to the graph by clicking the *Show Numbers* icon. Record the abundance values for the population under the 3.9% harvesting regime in Table 8.4.
17. Under the <u>R</u>esults menu select <u>H</u>arvest Summary. View the weights of harvested tuna by clicking on the *Show Numbers* icon in the upper left corner of the window. Record the kilograms of tuna harvested at each time step in Table 8.5.

Table 8.5. Amount of tuna harvested under 2 different harvesting rates.

	Amount of Tuna Harvested (Kg)	
Time step	1% Harvest	3.9% Harvest
1		
10		
20		
30		
40		
Total (kg)		

Questions

1. How long will it take for the population to double with no harvesting (in years and time steps)? How does this compare to the doubling time for *Bluefin1.st* with no harvesting?

2. How long will it take the population to double with a harvesting rate of 1%? With a harvesting rate of 3.9%?

3. How do the yields at the fortieth time step compare for the 2 harvesting rates?

4. Based on these results, what argument would you use to convince someone that the maximum sustainable yield was the best strategy for people to adopt?

5. If you were to add environmental stochasticity, do you think you would come to the same conclusions? In particular, do you think the optimal harvest rate would be the same?

6. Discuss some of the conflicts that are involved in imposing stricter catch limits on a species such as the bluefin tuna. Include an evaluation of the short-term versus long-term benefits and costs.

Your lab report should include the following:

1. One completed column of the market harvest table (Table 8.1) for African animals, 1 bar graph, and answers to questions 1 and 2 for Part 1 of Exercise A
2. Completed Table 8.1 and answers to questions 1 through 4 for Part 2 of Exercise A
3. Two trajectory graphs and 1 completed abundance table (Table 8.2) for Part 1 of Exercise B
4. One completed harvest table (Table 8.3) and answers to questions 1 through 7 for Part 2 of Exercise B
5. Three trajectory graphs, 1 completed abundance table (Table 8.4), 1 completed harvest table (Table 8.5), and answers to questions 1 through 6 for Part 3 of Exercise B

References

Fa, J., J. Juste, J. Perez del Val, and J. Castroviejo. 1995. Impact of market hunting on mammal species in Equatorial Guinea. *Conservation Biology*, 9(5): 1107–1115.

Ganzhorn, J. U., S. Sommer, J. P. Abraham, M. Ade, B. M. Raharivololona, E. R. Rakotovao, C. Rakotondrasoa, and R. Randriamarosoa. 1996. Mammals of the Kirindy Forest with special emphasis on *Hypogeomys antimena* and the effects of logging on the small mammal fauna. *Primate Report*, 46-1: 215–232.

Hilborn, R. and C. J. Walters. 1992. *Quantitative Fisheries Stock Assesment*. Chapman and Hall: NY.

International Commission for the Conservation of Atlantic Tuna. 1997. *Statistical Bulletin*, Madrid.

Robinson, J. and K. Redford. 1991. Sustainable harvest of Neotropical forest animals. In J. Robinson and K. Redford (eds.), *Neotropical Wildlife Use and Conservation*, (pp. 415–429). Chicago University Press, Chicago, IL.

Laboratory 9
The African White Rhino: Too Many for Their Own Good?

Once a park or reserve is created, should we take a hands-off approach to managing populations? If not, how much should we actively regulate with populations to conserve them? There have been many disagreements between conservation organizations and park managers on these issues. Anyone entering the fields of environmental studies or conservation biology should be aware of the debate behind this issue as well as the ecology behind these ideas.

Introduction

At the center of many disagreements over management strategies are two conflicting approaches: conservation versus preservation. These terms are often used interchangeably but in fact have very different meanings. Conservation involves the deliberate and planned management of wildlife or natural resources. Preservation involves setting aside a natural area and strictly protecting it from human activities. Although conservation may involve active interference or regulation, preservation involves restricting all human involvement or impact.

Minimizing our impact on natural areas is often desired, but active interference to manage populations is sometimes necessary if we want to conserve them. Humans in some way affect almost every natural area. Such things as air and water pollution often affect reserves that are protected from habitat destruction or harvesting of plants and animals. Populations also may be affected simply by the limited size of the reserve. Small reserve size may limit migration, dispersal, and the number of individuals the reserve or park can support over a long period of time. All of these factors are potential hazards to long-term persistence of populations.

Limiting Population Growth

It is often necessary to manage populations that are quite large and face no danger of extinction. The most controversial aspect of active management involves limiting the population size of such a species. **Culling** is the practice of removing individuals from an already overcrowded population (often through hunting or trapping). This lowers the density of a population by increasing mortality. Another alternative is to **sterilize** males or females so that they are unable to reproduce. Sterilization lowers the population's growth rate by lowering fecundity. Alternatively, some individuals may be **translocated**, or moved to another population. Translocation is a way of artificially increasing emigration rates. All of these practices have the effect of reducing a managed population's size or growth rate. It is important to understand the pros and cons of limiting the size of a population. There are a number of circumstances that might warrant active management of a population to protect it or other species with which it interacts.

Predation

One situation in which some people find it necessary to manage a population is if it has a negative impact on another species. For example, a predatory species may be severely reducing the population size of its prey. Active management of the predator population may be carried out because the prey is being threatened by extinction or because humans themselves benefit from the prey species. The caribou and wolves of Alberta, Canada serve as a good illustration of a predator-prey relationship that may be out of balance. Local populations of caribou are declining and being threatened by extinction. The wolf, which preys on caribou, has been blamed for their decline. Because the wolf also feeds on other animals, the low caribou abundance has not limited the size of the healthy wolf population. To preserve the caribou, a wolf control program has been recommend by local interest groups.

Exotic species

Active management may also be necessary to control exotic species populations when they threaten native species composition. An exotic species (also called a nonnative or invasive species) is one that moves into an area that is not its natural habitat. While most exotic species are either unsuccessful or cause no harm to native species, some may be a particular concern to native communities. Successful invasives may cause native species extinction through competition or predation. In the United States, exotic species invasions are the number one cause of endangerment of species listed as threatened or endangered by the U.S. Fish and Wildlife Service (Czech and Krausman 1997).

Humans may introduce species intentionally, as in the case of ornamental plants, or unintentionally, like some insect pests that have been shipped with crops from other areas. Humans may also indirectly permit the invasion of a nonnative species by disturbing a habitat and making it more suitable for nonnative species. An example of an intentional introduction gone awry is that of mongooses to many islands off of Central and South America. The animals were introduced to these areas to reduce the number of poisonous snakes. The mongooses have now begun to threaten the native populations of harmless snakes, amphibians, and birds. A plan to reduce the mongoose population on these islands has been proposed.

Human interests

People may also choose to actively control populations is when a particular species threatens human life or livelihood. As human settlements begin to encroach on undeveloped wilderness areas, the incidence of human interactions with large carnivores increases. Killing animals out of fear of personal harm, or harm to one's livestock, has lead to the decline of many of the world's large carnivore populations. The harvesting of jaguars in South America is one example.

Threats to peoples's way of life can also spawn controversy over conservation issues. In some rivers in the Rocky Mountains of Colorado, game species such as brown trout, rainbow trout and smallmouth bass have been introduced for recreational fishing. Where these nonnative species become established (and in some cases have introduced new diseases), native species such as the cutthroat trout rapidly decline in abundance and may become extinct in that river. There are strong political pressures to maintain the game species despite the possible extinction of some of the native species.

Overpopulation

Many scientists believe that organisms in dense, crowded populations may be too numerous for their own good. Individuals in populations that are held below the carrying capacity are often healthier, may compete less for food, and are better able to survive seasonal food shortages. These arguments are often given for the culling of herbivores such as rabbits and deer.

In some cases a population may overshoot its carrying capacity and begin to affect the environment irreversibly. In this case, the individuals become so numerous that they deplete their resources, causing the community dynamics change. Erosion, loss of other species, and the spread of disease may occur. Culling or translocation of individuals may be necessary for the future survival of both its own population and that of other species in the community.

Even though a species might be rare on the global scale, it may be locally abundant, and even above its carrying capacity, within a park or reserve. Should such a population be left alone, or should park managers actively control the population by culling, sterilizing, or moving individuals?

Elephants of Zimbabwe

One case where the conservation versus preservation debate is quite active involves the elephant population in Zimbabwe (Figure 9.1). Elephant populations in Africa have been reduced to very low numbers in the past from ivory hunters and diseases such as rinderpest. Since the increasing awareness of the declining populations, protected areas have been established and hunting has been made illegal in many areas. The sudden decrease in mortality resulting from their protection has caused the elephant population in Zimbabwe to rise dramatically despite the shrinking habitat and ever-growing human population.

Figure 9.1. An African elephant.

The elephant population in Zimbabwe has been increasing at a rate of 5% per year. That means that at the present growth rate, 2½ as many elephants will be present in 20 years. Some people believe that the populations will naturally stabilize to an acceptable level. Others have argued that the present growth rate will actually have a negative effect on the population in the future. Many people feel that it is wrong to kill the elephants to stabilize this growth trend because they are still rare on the global level.

Fences border many of the reserves to prevent the elephants from trampling crops or transmitting diseases to cattle. The elephants that would normally have dispersed to other areas to avoid high densities or crowding cannot leave. The parks in Zimbabwe cannot support such large numbers of elephants and the trees and vegetation that support them are being eaten or destroyed. This loss of vegetation has caused erosion of the soil so that the area can no longer support the same kinds of plants. This causes the community dynamics to change, and populations of other species become threatened. Although controversial, the regular and controlled culling of elephants has been occurring In Zimbabwe since the 1980s.

Exercise A: White rhinos at Umfolozi Reserve

Background

The rhinoceroses of Africa and Asia have a history similar to the elephant. Conversion of their natural woodland and grassland habitat for agriculture and development, along with hunting pressures for their highly valued horns, has taken a toll on all 5 species of rhinoceros. In the 1990s only about 11000 individuals of all 5 species are left.

The southern white rhino (*Ceratotherium simum*) came very close to extinction by the beginning of the twentieth century. Although its range was greatly reduced from its former distribution, recent protection of habitat and bans on hunting have caused the population to increase from 20 or 30 individuals to about 6000. Within the Umfolozi reserve in northeastern South Africa, the white rhino population is increasing at a rate of 9.5% per year.

Does this mean the white rhinoceros is finally doing well and it should be left alone? Although many conservationists may look at this sudden population explosion as beneficial for this species, it may not be good for the population in the long term. The population is at nearly twice the estimated carrying capacity. If the park was surrounded by forest where the animals could disperse, this might not be a problem. A fence, though, surrounds the reserve and the surplus animals must stay within the protected area. The rhinos feed almost exclusively on grasses, and their high numbers are causing deterioration of the habitat. There are changes occurring in the vegetation of the area where these large animals forage. Ecologists believe that overgrazing is creating a new but less productive grassland (i.e., one that will support fewer animals).

Themeda grasslands are naturally occurring in the park, yet much of the grass has been grazed almost to the ground and has had little chance to grow back before being eaten again. When the nutrient reserves in the roots of the plants are used up, they cannot get enough sun to photosynthesize and so the plants die. As the surface of the ground is defoliated (plants are removed), nothing is present to hold on to the rich organic material in the soil. Natural forces such as water and wind begin to erode the rich exposed soil, leaving a sandy soil with little nutrients. Not only is the species composition different, but plants that grow back are not as vigorous because the soil has changed. With the loss of vegetation, the area is becoming drier. Springs have begun to dry up and the flow of rivers has been reduced.

In this exercise you will run some computer simulations to evaluate how the white rhino population of the Umfolozi reserve should be managed. Table 9.1 lists the life history characteristics for 7 stages of white rhinos in the Umfolozi Reserve.

Table 9.1. Life table of white rhinos of the Umfolozi Reserve (modified from Owen-Smith 1981).

Age (years)	Abundance	Fecundity	Survival Rate
0–7	489	0.00	0.83
7–14	242	1.60	0.90
14–21	130	1.60	0.90
21–28	70	1.60	0.90
28–35	38	1.60	0.90
35–42	20	1.60	0.90
42–49	11	1.30	0.00

Exercises

1. Open the program RAMAS EcoLab (double-click on its icon with your mouse).
2. Click on *Age and Stage Structure*.
3. Under the *Model* menu select *General Information*. In that window, title this file White rhinoceroses.
4. Also in the *General Information* window, change the *Duration* from 0 to 10. By doing this, you have set the program to simulate ten time steps of reproduction. In this exercise, each time step is 7 years long.
5. Also in the *General Information* window, set the number of *Replications* to 1. You will run one simulation to view how the size of the rhinoceros population is expected to change over time. Click OK to exit this window.
6. Under the *Model* menu select *Stages*. Click the *Add* button until you have 7 stages. Under *Name* enter the age classes found in Table 9.1. Make no changes to the *Average Weight* column. Click OK to exit this window.
7. Under the *Model* menu select *Initial Abundances*. The age classes you entered in the previous step will appear in the window. Under each age class, enter the appropriate number of individuals as listed in Table 9.1.
8. Under the *Model* menu select *Stage Matrix*. The fecundities and survival rates are listed in the Table 9.1. Using these values, set up the matrix as you learned in Laboratory 4. Click OK to exit this window.
9. Now you are going to run the simulation; under the *Simulation* menu select *Run*, or press *Ctrl-R* on your keyboard. At the bottom right corner of this window you will see a message when the simulation is complete. At this point you may close the window by clicking on the X in the upper right corner of the window.
10. Under the *Results* menu, select *Trajectory Summary*. Here you can view the population growth curve. The time scale, like the age classes, is in seven-year time steps. Print the plot of the *Trajectory Summary* by clicking on the printer icon.
11. You can view the numbers corresponding to the plot by clicking the *Show Numbers* button in the upper left corner of the *Trajectory Summary* window. You can return to the plot by clicking the *Show Plot* button, also in the upper left corner of the *Trajectory Summary* window.

12. In Table 9.2, record the *Average Abundance* of the population at time steps 0, 5, and 10.
13. Exit the *Trajectory Summary* by clicking on the X in the upper right corner of the window.

Table 9.2 Average abundance of the rhinoceros population.

Time Step	Average Abundance
0	
5	
10	

Questions

1. How many year(s) does each time step represent?

2. How many time steps (and years) did it take the unmanaged population to double in size?

3. If you were to add demographic and environmental stochasticity to this model, how would this affect the prediction about the way the population size will change in the next 100 years?

4. What consequences do you think a further increase in population size will have on the reserve's ecosystem?

Exercise B: Management of an overpopulated reserve

Background

The white rhinos are estimated to be currently at twice the carrying capacity of the Umfolozi Reserve. It is feared that if the population continues to increase or even remain at its present density, the ecosystem will be irreversibly changed and will not be able to support rhinoceroses at all. In this exercise you are asked to make recommendations for controlling this exponentially increasing population. You will use computer modeling to come up with a management regime for controlling this population through either culling or translocation.

Exercises

The first step is to bring the population size down to carrying capacity so that no further damage to the ecosystem occurs. To accomplish this, park officials may either cull or translocate 50% of the animals. Then, to maintain this population size, park officials will remove a certain percentage of the population in each time step. You will modify the file you created in the previous exercise.

1. Under the *Model* menu select *Initial Abundances*. The age classes you entered in the previous exercise will appear in the window. Under each age class, enter half the number of individuals as listed in Table 9.1. If the original abundance is odd, round the new value up to the next highest integer. Click OK to exit this window.
2. Now you are going to run the simulation without any additional management practices such as removal of additional animals at each time step. Under the *Simulation* menu select *Run*, or press Ctrl-R on your keyboard. Close the window by clicking on the X in the upper right corner of the window.
3. Under the *Results* menu, select *Trajectory Summary*. Here you can view the population growth curve. The time scale, like the age classes, is in seven-year time steps.
4. You can view the numbers corresponding to the plot by clicking the *Show Numbers* button in the upper left corner of the *Trajectory Summary* window. You can return to the plot by clicking the *Show Plot* button also in the upper left corner of the *Trajectory Summary* window.
5. In Table 9.3, record the *Average Abundance* of the population at time steps 0, 5, and 10 under the column indicating a harvest level of 0.00. Exit the *Trajectory Summary* by clicking on the X in the upper right corner of the window.

122   Laboratory 9

Even though the population was brought down to its carrying capacity; it still has a positive growth rate, therefore, the next step is to determine the proportion of adults that need to be removed every time step to maintain a 0 growth population.

6. Under the *Model* menu select *Management & Migration*. Under *Management Action* click Add and select *Harvest/Emigration* by clicking on the circle next to that phrase. Under *Quantity* select *Proportion of Individuals* by clicking on the circle next to that phrase.
7. Decide what proportion of the population you would like to harvest. If you think that harvesting 20% of the population will sufficiently lower the growth rate, then enter 0.20. Using the two pull-down menus at the bottom of the window, apply this action to stages 8–14 through 42–49.
8. Run the simulation and record the average abundance of the population at time steps 0, 5, and 10 in the appropriate column of Table 9.3.
9. In the *Management & Migration* window, enter other harvest rates (values from 0 to 1.0) one at a time until you find one that stabilizes the population. Remember that, to be considered stable, the population should remain at about the same abundance for about five time steps. Record the results of each successive trial in Table 9.3.
10. When you have finished, print out a trajectory graph of the harvest rate that stabilized the population by clicking on the printer icon.
11. Review a summary graph of your management action by selecting *Harvest Summary* found under the *Results* menu. Print this graph as well.

Table 9.3. Average abundance of rhinos under different harvesting regimes.

	Average Abundance of White Rhinos						
Harvest Level:	0.00						
0 Time Steps							
5 Time Steps							
10 Time Steps							

Questions

1. What is the proportion of adult animals in each age class to be removed every time step to balance the rhino population at Umfolozi Reserve?

2. Approximately how many animals are to be removed every time step to balance the rhino population at Umfolozi Reserve?

3. Discuss some reasons you think people object to culling animals in a protected area.

4. Why is it important that culled populations be monitored very carefully and on a regular basis?

Exercise C: Balancing a declining population

Background

Removing animals seems to be a necessary management strategy to conserve rhino populations in areas like the Umfolozi Reserve. In other areas, though, the white rhino populations are declining because of illegal poaching combined with natural pressures such as hyena predation on young. Is it possible to balance these declining populations by translocating animals from overpopulated areas? In this exercise you will attempt to balance a declining population by introducing new animals to the area. You will also consider the cost of translocation versus other management strategies.

Below is a life table from a hypothetical declining white rhino population (Table 9.4). These data are based on realistic survival rates and fecundity values experienced by rhinoceroses in declining populations. You will use these data in the following exercises.

Table 9.4. Life table of a declining white rhino population.

Age (years)	Abundance	Fecundity	Survival Rate
0–7	86	0.00	0.15
7–14	21	0.00	0.50
14–21	17	1.50	0.30
21–28	14	1.50	0.30
28–35	7	1.50	0.30
35–42	3	1.50	0.30
42–49	2	1.00	0.00

Exercises

1. Open a new file in the *Age and Stage Structure* window of RAMAS EcoLab by clicking on the *New* file icon in the upper left corner of the window. You will be asked if you want to save the file from the previous exercise, but you need not do so.
2. The *General Information* window from the *Model* menu will automatically open. In that window, title this file *White Rhinoceros Translocation*.
3. Also in the *General Information* window, change the *Duration* from 0 to 10. By doing this, you have set the program to simulate ten time steps of reproduction. In this exercise, each time step is seven years long.
4. Also in the *General Information* window, set the number of *Replications* to 1. You will run one simulation to view how the size of the declining rhinoceros population is expected to change over time. Click OK to exit this window.
5. Under the *Model* menu select *Stages*. Click the *Add* button until you have 7 stages. Under *Name*, press the age classes found in Table 9.4. Make no changes to the *Average Weight* column. Click OK to exit this window.
6. Under the *Model* menu select *Stage Matrix*. The fecundities and survival rates are listed in Table 9.4 for the declining rhino population. Using these values, set up the matrix as you learned in Laboratory 4. Click OK to exit this window.
7. Now you are going to run the simulation; under the *Simulation* menu select *Run*, or press *Ctrl-R* on your keyboard. At the bottom right corner of this window you will see a message when the simulation is complete. At this point you may close the window by clicking on the X in the upper right corner of the window.
8. Under the *Results* menu, select *Trajectory Summary*. Here you can view the population growth curve. The time scale, like the age classes, is in seven-year time steps.
9. You can view the numbers corresponding to the plot by clicking the *Show Numbers* button in the upper left corner of the *Trajectory Summary* window. You can return to the plot by clicking the *Show Plot* button, also in the upper left corner of the *Trajectory Summary* window.
10. In Table 9.5, record the *Average Abundance* of the population at time steps 0, 5, and 10. Exit the *Trajectory Summary* by clicking on the X in the upper right corner of the window.

In the next step, your goal is to try to stabilize this declining population by introducing new individuals into the park.

11. Under the *Model* menu select *Management & Migration*. Under *Management Action* click *Add* and select *Introduction/Immigration* by clicking on the circle next to that phrase. Under *Quantity* select *Number of Individuals* by clicking on the circle next to that phrase.
12. Decide the number of adults you would like to add to the population to stabilize it. Because the population is very small to begin with, the addition of only a few individuals may have a significant effect. Using the two pull-down menus at the bottom of the window, apply this action to stages 8–14 through 42–49.
13. Run the simulation and record the average abundance of the population at time steps 0, 5, and 10 in the appropriate column of Table 9.5.
14. In the *Management & Migration* window, enter other introduction rates (they must be integers) one at a time until you find one that stabilizes the population. Remember that, to be considered stable, the population should remain at about the same

abundance for about five time steps. Record the results of each successive trial in Table 9.5.
15. When you have finished, print out a trajectory graph of the introduction rate that stabilized the population by clicking on the printer icon.

Table 9.5. Average abundance of rhinos with different introduction rates.

	Average Abundance of White Rhinos							
Number Introduced:	0							
0 Time Steps								
5 Time Steps								
10 Time Steps								

Questions

1. In the declining population, how many animals had to be introduced *per age step* every year to stabilize the population? Can this number be taken from the Umfolozi Reserve without hurting the population (look at your harvest summary information)?

2. If it costs $30,000 (this is a conservative estimate) to translocate 1 individual, how much will it cost each year to balance this population?

3. Because conservation money cannot be guaranteed forever, the practice of translocating animals from 1 population to another will eventually cease. What do you think will happen to the declining population when this happens? Will the population be able to maintain constant numbers without the addition of animals at each time step?

4. Can you think of some alternative conservation actions that might help this population recover that might be a more efficient and sustainable use of the money available?

5. Translocating animals might not be a sustainable practice for rhinos, yet there are other species and other populations for which it is very effective. What characteristics would you look for in a species or in a population to decide that translocation might be an effective conservation strategy?

Your lab report should include the following:

1. An abundance table (Table 9.2), population trajectory graph, and answers to questions 1 through 4 for Exercise A
2. An abundance table (Table 9.3), population trajectory graph, a harvest summary graph, and answers to questions 1 through 4 for Exercise B
3. An abundance table (Table 9.5), population trajectory graph, and answers to questions 1 through 5 for Exercise C

Reference

Owen-Smith, N. 1981. The white rhino overpopulation problem and a proposed solution. In Jewell, P., S. Holt, and D. Hart (eds.), *Problems in Management of Locally Abundant Wild Mammals*, (pp.129–150). Academic Press, NY.

Czech, B. and P. R. Krausman. 1997. Distribution and causation of species endangerment in the United States. *Science* 227: 1116–1117.

Laboratory 10
The Wild Ass and the Black Footed Ferret: Reintroduction of Endangered Species

Variation is always present in the natural world. To simplify analyses, we often assume this variation does not exist or is not important. Is this reasonable? What happens to the predictions when there is error in the original data? What are the limitations of the predictions? In the following laboratory you will look at how variation affects populations and the predictions we make about the success of endangered species.

Introduction

Can you recall a weather forecast predicting a clear sunny day that turned out to be cold and rainy? Have you noticed that forecasts are less accurate the further in the future they try to predict? Suppose you are planning a Saturday barbecue. On Tuesday the 5-day forecast predicts fair weather for the weekend. By Friday, rain is predicted for Sunday, but not for Saturday. Sure enough, you wake on Saturday morning to light drizzle that turns into a downpour by noon. What happened?

Time and Uncertainty

The barbecue story demonstrates a very important aspect of probability. As we predict events farther into the future, the certainty of any particular outcome decreases. For example, you would not have much confidence in a weather forecast for the first of June of next year. The predictions for the day after tomorrow are based on the prediction for tomorrow. If we are not completely certain about tomorrow, we are even less certain about the day after that.

Uncertainty is also important in conservation biology. If we are not sure what the population size will be in 5 years, we are less certain what the population size will be in 15. We can make an educated guess from patterns, data, and experience. The more realistic the model is, the better the predictions are. These predictions, though, are only probability statements of what is to come. It is critical to understand the limitations of prediction and probability. When dealing with a lot of variation in the natural world, it is unreasonable to expect the predictions for what will happen in 200 years or even in 100 years to be completely accurate.

Predictions and Error

We have little confidence in predictions made far into the future; we also have little confidence in predictions made from data that is incomplete or inaccurate. Suppose the estimate initial population abundance is 80 instead of the actual 100, the prediction will be inaccurate. If we assume 10 females who are not breeding are actually males, we will inflate the estimate of fecundity for the entire population, because the total number of offspring will be attributed to fewer females than actually exist in the population. Similarly, if we neglect to count several juveniles that were counted when they were infants, we will underestimate the survival rate of infants living to be juveniles. In the previous laboratories we have used RAMAS software to predict the future abundances and growth rates of a population. When we made the predictions, we assumed we had accurate measurements of the population size, and birth and death rates. What happens when this information is incorrect or variable?

Environmental Stochasticity

In several of the previous exercises the models included demographic stochasticity, or random variation in birth and death rates. However, they did not include environmental stochasticity, or random variation in the environment. The environment varies a great deal over time. Some of this variation is predictable. In the temperate zones, such as most of the United States, seasons are a predictable variation in weather. We can assume that the temperature will be cooler from December to February than from June through August. Even long-term events may be predictable. El Niño is a long-term cycle in the weather. Californians can expect wet weather every few years when El Niño water currents cycle in the southern Pacific Ocean.

Much of the variation in the weather, however, cannot be predicted. For example, although the weather patterns from El Niño were predicted during the winter of 1997–1998, the intensity and duration of heavy rainfall in California was not certain. When there are extreme weather patterns, both animal and plant populations can be affected. Survival rates of individuals can drop drastically during a particularly harsh winter. Other dramatic events in the environment may also affect populations. Droughts, flooding, and fires, among other natural events, can cause considerable disruption in a population. These extreme events are called catastrophes. Because they are so rare, we cannot with certainty predict when, if ever, they will occur near the population, and it is difficult to realistically incorporate them into a model.

It is critical to be careful about when and how data are collected from the population. When conditions are particularly favorable, for example, after a mild and wet winter, birth rates may be unusually high and death rates unusually low. The converse is usually true from bad years. If we have data from a long period of time, we can incorporate the natural variation in birth and death rates into the models. For most models, we are interested in the natural year-to-year variation in survival rates and fecundities.

If we do not have a lot of information on a population, we need to be especially cautious about the predictions we make. One strategy is to use data from different populations of the same species to estimate of the amount of variation that is possible in the species. Careful measurements can be made of the resources available to the population for a rough assessment of the quality of the environment that year. Whenever we use data from a population, we need to be aware of the assumptions we are making about the population. As an example, it is usually assumed that the measurements represent average values for population parameters, but often we know little or nothing about the conditions under which the data were collected.

One of the important ways that environmental stochasticity can be incorporated into a model is from long-term monitoring of populations. If a species is carefully monitored for a long time, say 30 or 40 years, it is likely the data will include much of the normal variation in environmental factors. Using average values for survival rates and fecundity values and a standard deviation around these averages simulates this type of variation. The standard deviation is a measure of the amount a certain parameter (e.g., survival rate) varies around the average. If there is a large standard deviation, there is a lot of variation, whereas a small standard deviation indicates little variation. When running a model with stochasticity, every time the model is run, the predictions will be different, because of random variation. Therefore, it is important to run the model as many times as possible (i.e., run as many replications of the simulation as possible) to see how much variability exists in the predictions.

Uncertainty and variation are central concepts in conservation biology. Simply put, uncertainty exists in the real world, so it should be incorporated it into simulations when modeling the future of an endangered species. For example, if we wanted to estimate how low a certain species' population size may become in some time interval, we would run a simulation with stochasticity incorporated several times (with many replications) and determine in how many of the replications the population size falls below some threshold.

If we are interested in the probability of a population falling below 100 individuals in the next 20 years, we would run a simulation for a duration of 20 years many times (say 1000), and determine in how many of the replications the population size falls below 100. If the population falls below 100 individuals in 200 of the replications, we can estimate that there is a 200/1000, or 20%, probability of the population falling below 100 in the next 20 years.

In the following sections you will learn about some types of management actions that are taken. Then, in the exercises, you will model these management actions via computer simulation, incorporating the effects of stochasticity.

Releasing Individuals of Endangered Species

For populations of endangered species that have dwindled to small numbers, or have become locally extinct, we can introduce (translocate) either captive-bred or wild-captured individuals to try to rescue or restore a population. There are 3 different approaches to trying to establish or stabilize a population, reintroduction, introduction, and augmentation. Reintroduction is the release of individuals into the historic range of the species, areas where a population once existed but is no longer present. Introduction is the release of animals into areas outside of the historic range of the species. Augmentation is the release of individuals into a small population to help boost the population size. The translocation of rhinos in the last laboratory is an example of augmentation. All 3 techniques are useful for expanding the size and range of an endangered species' distribution. However, release programs have proven difficult, problematic, and expensive.

Griffith et al. (1989) conducted a detailed study of 198 programs for bird and mammal establishment programs (i.e., reintroduction, translocation, and augmentation). They found that the success of the program was dependent on a number of factors. Figure 10.1 demonstrates the success rates of reintroduction programs that differ in some of these factors. It was much easier to establish game species than threatened, endangered, or sensitive species. The chance of success was much greater when the release occurred in excellent-quality habitat rather than poor-quality habitat. Wild-caught animals were more likely to successfully establish themselves than captive-reared animals. These success rates suggest that there are several complications in successfully reintroducing a species.

Why would animals raised in captivity perform poorly when introduced back into their natural habitat? Individuals born in captivity generally do not have to work very hard to find food, escape predators, or find mates. In a zoo or a breeding facility, human caretakers provide both food and protection for the animals. Once these animals are released into the wild, they need to be able to care for themselves. Many long-lived animals, especially mammals and birds, pass through an intense learning stage where adult animals teach the young how to find resources, avoid danger, and when and where to migrate. If we plan to release naive animals, we need to provide them with the tools to survive.

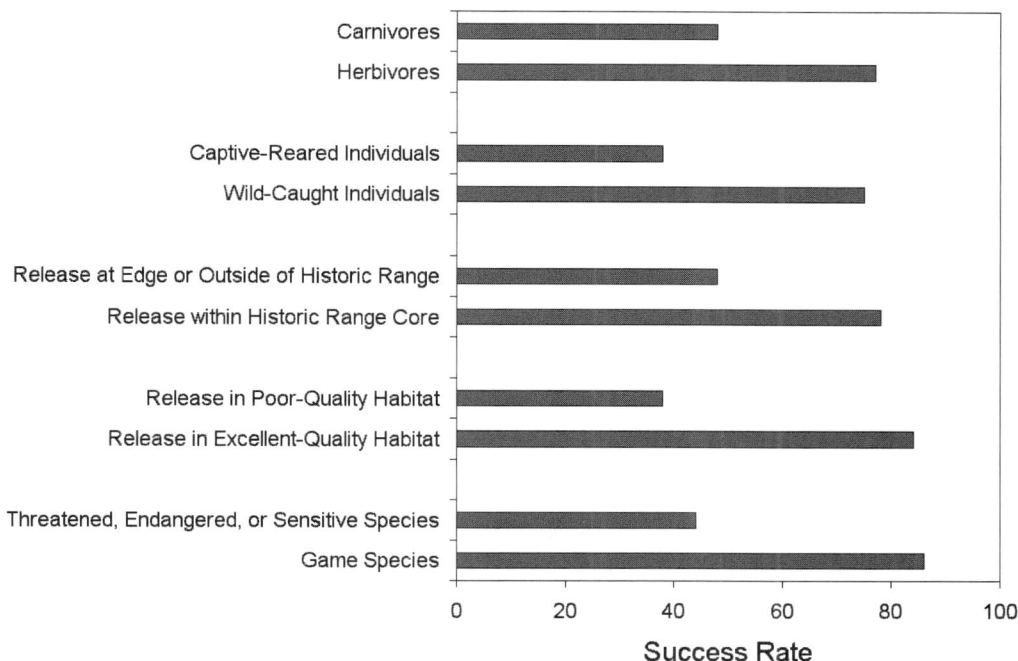

Figure 10.1. Comparison of factors that affect the success rate of establishment programs (adapted from Griffith et al. 1989).

The whooping crane is a well-known example of an endangered species whose complex behavior and biology have reduced the success of reintroduction programs. One of the key sets of behaviors that are socially transferred in these cranes is parenting skills. Whooping cranes apparently learn how to care for their offspring by interacting with other members of the population. Captive-reared animals are at a disadvantage because they are usually raised by humans directly from eggs and do not have experience interacting with adult whooping cranes. A number of programs have been implemented to try to teach the young cranes necessary social skills. One such program includes "adoption" by sandhill cranes, with moderate success.

An additional problem unique to releasing social animals is acceptance of the released individuals into a normal social group. In many social animals, a lot of aggression is directed at strangers or unknown individuals trying to gain access to the group. When individuals join a group, they often spend a period living on the edge of the group, a dangerous position for an inexperienced animal. One way to avoid this problem is to release several individuals at the same time to make a new social group. Unfortunately, in this case none of the individuals know the territory or where the necessary resources are.

If enough research has been conducted on the species of interest, we can often predict problems that individuals will have with their environment and be able to address these problems before we introduce them to a hostile situation. A small monkey from South America, the golden lion tamarin, has been the focus of a long-term reintroduction program. Because the foods normally eaten by monkeys in the wild are diverse and often difficult to handle, primatologists (those who study monkeys and other primates) are devising ways to teach captive individuals to forage as they would in the wild.

Different species have very different requirements. As Figure 10.1 illustrates, it is much easier to successfully reintroduce an herbivore or plant-eating species than a predator species. Predators such as wolves and cats present another obstacle. These animals must hunt and kill the appropriate prey items to survive. Wolves regularly take down medium to large mammals and cannot do so without learning through trial and error. Released individuals must be able to locate and capture prey to survive.

One well-known example of the complexities of reintroducing a predator is the California condor. Condors are large scavengers and need a lot of open space. The last free-living condor was estimated to have foraged over 2.4 million hectares (5.93 million acres) of habitat. The reintroduction program has faced many problems, not the least of which is teaching the birds to avoid high-voltage electricity wires. Although captive breeding programs have been quite successful, it is unclear whether the few individuals that have been released into the wild will learn the necessary skills to survive and reproduce.

Establishing a population can be more difficult than maintaining it. After a minimal number of individuals are able to survive and produce offspring, the population growth rate should increase, once the wild-born offspring mature. Even after a single generation, wild-born individuals tend to be more competent than their captive-bred parents are. Although some reintroductions are doomed to failure, often this is because the habitat has been degraded or is too small to maintain a viable population, or the factors that caused the reduction or extinction are still present. Adequate training and knowledge will help reduce unnecessary reintroduction failures.

A final consideration and challenge of release programs is their high cost. Millions of dollars can be spent on a single population. Often the long-term viability of that population is uncertain. Conservation biologists, as well as the public, face tough decisions about how to allocate limited funds. Individuals and groups that are not sympathetic toward conservation efforts can grow to resent failed efforts and what they may consider "wasted" money.

Exercise A: The Asiatic wild ass

Background

The Asiatic wild ass (*Equus hemionus*) became locally extinct in the Middle East in the early part of the twentieth century. The last wild individual was sighted in Syria in 1938. In the 1980s, a population was reintroduced into Israel. Between the years 1983 and 1987 fourteen females and nine males were released from a captive breeding population into the Makhtesh Ramon Reserve (Saltz and Rubenstein 1995). Overall, the survival of the released animals was fairly high, except for several older females who died soon after being released. But by 1993, the herd had only expanded from fourteen to sixteen females. Populations with such a slow growth rate are at risk of declining for quite a long time due to stochastic effects.

The researchers who conducted the reintroduction wanted to know why the population was growing so slowly. They noticed several interesting things. First, the longer a female had been in the wild, the higher her reproductive rate was. Second, they noticed that wild-born daughters had much higher reproductive rates than their captive-born mothers did. Third, older females had higher reproductive rates than younger ones. All three points suggest that the population growth rate should increase over time as more individuals are wild-born, and these wild-born daughters mature.

An important lesson from this study is that it is important to follow populations for some time after individuals have been released. In the following exercise, you will look at projections for the Israel population using reproductive values from only the first four years after the individuals were released. You will then add the information about the reproductive rates of the wild-born daughters to see how this will change the predictions for the viability of the population.

Exercises

Part 1: The first four years

1. Open the program RAMAS EcoLab (double-click on its icon with your mouse).
2. Click on *Age and Stage Structure*.
3. Under the *File* menu, select *Open* and choose *Wildass.st*. This file contains an age-structured matrix of released captive-born individuals.
4. Under the *Model* menu select *General Information*. Notice that the *Duration* is set at 20-5. Which means that the program is set to simulate 20-5 time steps of reproduction. In this exercise, each time step is three years long.
5. Also in the *General Information* window, notice that the *Number of Replications* is set at one. You will run one simulation at a time to view how the size of the reintroduced wild-ass population is expected to change over time. You will be including demographic stochasticity in your model. Click *OK* to exit this window.
6. Under the *Model* menu select *Stage Matrix*. Inspect the matrix and note that this is a six-stage matrix with three-year age classes. Notice also that the fecundities are higher for the older age classes, as described in the introduction of this exercise. Click *OK* to exit this window.
7. Now you are going to run the simulation. Under the *Simulation* menu select *Run* or press *Ctrl-R* on your keyboard. At the bottom right corner of this window you will see a message when the simulation is complete. At this point you may close the window by clicking on the X in the upper right corner.
8. Under the *Results* menu, select *Trajectory Summary*. Here you can view the population growth curve. The time scale, like the age classes, is in three-year time steps. Print the plot of the *Trajectory Summary* by clicking on the printer icon.
9. You can view the numbers corresponding to the plot by clicking the *Show Numbers* button in the upper left corner of the *Trajectory Summary* window. Record the final abundance in Table 10.1.
10. You can return to the plot by clicking the *Show Plot* button, also in the upper left corner of the *Trajectory Summary* window. Exit the *Trajectory Summary* window by clicking on the X in the upper right corner of the window.
11. Repeat steps 7 through 9 for a total of 10 times. Each time you run the simulation you are running another replication that includes demographic stochasticity. Record the final abundances each time you run the simulation in Table 10.1.

Table 10.1. Final abundance for 10 replications, with demographic stochasticity.

Replication	Final abundance
1	
2	
3	
4	
5	
6	
7	
8	
9	
10	
Average	

Questions

1. What does your table of final abundance values suggest about the effect of demographic stochasticity on projected population abundance?

2. Would you feel confident making management decisions for this population based on a single replication of the model? Explain.

Part 2: The next generation

1. Under the <u>M</u>odel menu select *Stage Matri<u>x</u>*. Modify the matrix to reflect the survival and fecundity of the next generation of wild-born daughters. Fill in the matrix using the values from Table 10.2. Be aware that you will not adjust the *Initial Abundances*. Click OK to exit this window.

Table 10.2. Life table for wild-born daughters of the Asiatic wild ass.

Age (years)	Fecundity	Survival Rate
0-2	0.08	0.86
3-5	1.04	0.77
6-8	0.44	0.73
9-11	0.90	0.73
12-14	1.40	0.54
15$^+$	0.50	0.00

2. Now you are going to run the simulation. Under the *Simulation* menu select *Run* or press *Ctrl-R* on your keyboard. At the bottom right corner of this window you will see a message when the simulation is complete. At this point you may close the window by clicking on the X in the upper right corner of the window.
3. Under the *Results* menu, select *Trajectory Summary*. Here you can view the population growth curve. The time scale, like the age classes, is in three-year time steps. Print the plot of the *Trajectory Summary* by clicking on the printer icon.
4. You can view the numbers corresponding to the plot by clicking the *Show Numbers* button in the upper left corner of the *Trajectory Summary* window. Record the final abundance in Table 10.3.
5. You can return to the plot by clicking the *Show Plot* button, also in the upper left corner of the *Trajectory Summary* window. Exit the *Trajectory Summary* window by clicking on the X in the upper right corner of the window.
6. Repeat steps 2 thorough 4 for a total of 10 times. Each time you run the simulation you are running another replication that includes demographic stochasticity. Record the final abundances in Table 10.3.

Table 10.3. Final abundance for 10 replications with new survival and fecundity rates.

Replication	Final Abundance
1	
2	
3	
4	
5	
6	
7	
8	
9	
10	
Average	

Questions

1. Do your predictions about the future growth rate of the population differ depending on whether you look at captive-born or wild-born females? How does this affect your predictions of the growth rate of the population from now on as compared to that from 1987 to the present?

2. Why do you think the reproductive rates of wild-born daughters may be higher than those of their mothers?

3. Did different characteristics of different individuals (e.g., age) affect the probability of survival? How could we incorporate this information into reintroduction programs?

4. How long would you recommend monitoring a newly released population before you could make realistic long-term predictions?

5. Does this exercise suggest that we should we give up on introductions if the population size is declining in the first years after the animals are released?

Part 3: Environmental stochasticity

Now you will examine the potential effects of random environmental variation on this newly introduced population. Assume that the survival and fecundity values that were measured during the first year after release are close to the average values that the population will experience. Normal variation in the climate, though, may change these values by as much as 20%. To realistically predict the long-term population abundance for the wild ass, you need to take this variation into account in the model. You will do this by adding a standard deviation to each survival and fecundity rate. For example, if the survival rate is 0.80 and you use a standard deviation of 0.10, most survival rates used by the model will fall between 0.80 − 0.10 = 0.70 and 0.80 + 0.10 = 0.90.

1. Under the *File* menu, select *Open* and choose *Wildass.st*. Recall that this file contains an age-structured matrix of released captive-born individuals. You will be asked if you want to save the file from the previous exercise, but you need not do so.
2. Under the *Model* menu select *General Information*. Change the *Number of Replications* from 1 to 50. You will run 50 individual simulations to view how the size of the reintroduced wild-ass population is expected to change over time. You will be including demographic stochasticity. Click *OK* to exit this window.
3. Now you are going to run the simulation without environmental variation as a standard by which to compare the effect of stochasticity. Under the *Simulation* menu select *Run*, or press *Ctrl-R* on your keyboard. At the bottom right corner of this window you will see a message when the simulation is complete. At this point you may close the window by clicking on the X in the upper right corner of the window.
4. Under the *Results* menu, select *Trajectory Summary*. Here you can view the population growth curve. The time scale, like the age classes, is in three-year time steps. Print the plot of the *Trajectory Summary* by clicking on the printer icon.
5. You can view the numbers corresponding to the plot by clicking the *Show Numbers* button in the upper left corner of the *Trajectory Summary* window. Record the average abundance with no environmental stochasticity in Table 10.4 for time steps 0, 5, 10, 15, 20, and 25.
6. You can return to the plot by clicking the *Show Plot* button, also in the upper left corner of the *Trajectory Summary* window. Exit the *Trajectory Summary* window by clicking the X in the upper right corner of the window.

Table 10.4. Average abundance with and without environmental stochasticity.

	Average Abundance	
Time	No Environmental Stochasticity	With Environmental Stochasticity
0		
5		
10		
15		
20		
25		

7. Now you will add the effect of environmental stochasticity. You will assume that the standard deviation of each fecundity and survival rate is 20% of the average value, where the average value is the number in the stage matrix. Calculate 20% of each value (multiply each value by 0.20) and enter these numbers in the appropriate place in Table 10.5.

Table 10.5. Standard deviations of survival and fecundity rates.

Stage	Mean		Standard Deviation	
	Survival	Fecundity	Survival	Fecundity
1	0.7629	0.00	0.153	0.000
2	0.7290	0.16		
3	0.7290	0.28		
4	0.7290	0.90		
5	0.5427	0.90		
6	0.0000	0.50		

8. Now you will enter the standard deviations you have just calculated into the model. Under the *Model* menu, select *Standard Deviation Matrix*. In the cell corresponding to each rate, enter the standard deviation for that rate. This matrix will resemble the stage matrix except that it will contain standard deviations rather than average values for each rate. Click *OK* to exit this window.

9. Now you are going to run the simulation. Under the *Simulation* menu, select *Run*, or press *Ctrl-R* on your keyboard. At the bottom right corner of this window you will see a message when the simulation is complete. At this point you may close the window by clicking the X in the upper right corner of the window.

10. Under the *Results* menu, select *Trajectory Summary*. Here you can view the population growth curve. The time scale, like the age classes, is in three-year time steps. Print the plot of the *Trajectory Summary* by clicking on the printer icon.

11. You can view the numbers corresponding to the plot by clicking the *Show Numbers* button in the upper left corner of the *Trajectory Summary* window. Record the average abundance with environmental stochasticity in Table 10.4 for time steps 0, 5, 10, 15, 20, and 25.

12. You can return to the plot by clicking the *Show Plot* button, also in the left corner of the *Trajectory Summary* window. Exit the *Trajectory Summary* window by clicking on the X in the upper right corner of the window.

13. Make a line graph of the average abundance with and without environmental variation. Place time on the x-axis and abundance on the y-axis.

Questions

1. Is the average abundance with environmental stochasticity higher, lower, or the same as the abundance without environmental stochasticity? Why?

2. How will adding environmental stochasticity affect the predictions about the long-term survival of the wild ass population?

3. Based on these results, would you recommend introducing more, fewer or the same number of individuals into the wild (assuming there are more available from other locations or captive breeding programs)? Explain.

Exercise B: The black-footed ferret

Background

The black-footed ferret (*Mustela nigripes*) is a popular symbol of the United States Endangered Species Act (Figure 10.2). Wild populations of the ferret declined until there were only twelve individuals left in the wild in 1986 (Seal 1989). All the individuals were brought into captivity because the risk of extinction in that single population was so high. The decision to capture the remaining individuals was controversial because of previous failed captive-breeding attempts. Fortunately, by 1987 scientists had a good understanding of the breeding biology and requirements of the ferrets.

Figure 10.2. The black-footed ferret.

What caused the dramatic decline of the black-footed ferret in the wild? Why were attempts ineffective at preventing the population from falling to such low levels? Two major factors reduced the abundance of the ferret populations. The first is that ferrets almost exclusively eat prairie dogs. The second is that ferrets are particularly vulnerable to the disease canine distemper.

Prairie dogs live in large colonies called towns. In the early twentieth century, prairie dog towns covered much of the prairie and could have supported a population of up to 800,000 ferrets. Ranchers believed, and still do, that prairie dogs interfere with ranching. Therefore, in the 1920s and 1930s, ranchers and wildlife managers began a campaign to exterminate prairie dogs. By the 1980s, almost 90% of the native prairie dog population had been eliminated. Black-footed ferret numbers declined with the decline of their prey species. Initial efforts in the 1970s to capture and breed ferrets failed.

In the early 1980s, only one ferret colony remained, in Meeteetse, Wyoming. A population study in 1984 indicated that the population was growing and producing a surplus of individuals. However, in 1985 the population had crashed to 58 individuals after an epidemic of sylvatic plague wiped out many of the prairie dogs. This was followed by an outbreak of canine distemper in the ferrets. In 1986, all twelve remaining ferrets were captured with the goal of eventually reestablishing at least two free-ranging populations in Wyoming.

The captive population increased to 349 ferrets by 1992. Reintroduction of ferrets to Shirley Basin, Wyoming are continuing, albeit slowly. In this exercise, we are going to examine long-term predictions about population growth rate. You will use the data from the Meeteetsee population at its peak to evaluate the viability of the Shirley Basin population.

Exercises

Sensitivity analysis

One important aspect of population viability analysis is that we know how much error in the estimates will affect the predictions. One way to answer this question is to perform a sensitivity analysis, much like the one performed in Laboratory 7. However, the question being asked here is quite different from the question in that lab. In that case, you were interested in determining how much the fecundity and survival rates of each age class contributed to the chances of survival of the population. Now you are interested in how much error in the estimates of these rates will affect the predictions.

1. Under the *File* menu, select *Open* and choose *Ferret.st*. This file contains an age structured matrix of the Meeteetse, WY population based on estimates of fecundity and survival when the population was growing and producing surplus individuals.
2. Under the *Model* menu select *General Information*. Notice that the Duration is set at fifty. Which means that the program is set to simulate fifty time steps of reproduction. In this exercise, each time step is 1 year long.
3. Also in the *General Information* window, notice that the *Number of Replications* is set at 25. You will run 25 simulations to view how the size of the ferret population is expected to change over time. You will be including demographic stochasticity. Click OK to exit this window.
4. Under the *Model* menu select *Stage Matrix*. Inspect the matrix, and note that this is a 4-stage matrix with 1 year age classes. The youngest age class is made up of nonreproducing juveniles. Click OK to exit this window.
5. Now you are going to run the simulation. Under the *Simulation* menu, select *Run*, or press *Ctrl-R* on your keyboard. At the bottom right corner of this window you will see a message when the simulation is complete. At this point you may close the window by clicking on the X in the upper right corner of the window.
6. Under the *Results* menu, select the *Trajectory Summary*. Here you can view the population growth curve. The time scale, like the age classes, is in 1 year time steps. Print the plot of the *Trajectory Summary* by clicking on the printer icon.
7. You can view the numbers corresponding to the plot by clicking on the *Show Numbers* button in the upper left corner of the *Trajectory Summary* window. Record the average population size at the time steps 0, 10, 20, 30, 40, and 50 in Table 10.6.
8. You can return to the plot by clicking the *Show Plot* button, also in the upper left corner of the *Trajectory Summary* window. Exit the *Trajectory Summary* by clicking on the X in the upper right corner of the window.

Table 10.6. Average population abundances under various conditions.

Time Step	Average Abundance		
	Original Values for Survival and Carrying Capacity	Juvenile Survival = 0.18	Carrying Capacity = 100
0			
10			
20			
30			
40			
50			

142 Laboratory 10

What if you overestimated the survival rate of the juveniles? How much would this affect your predictions for the future growth of this population? To answer this question you will decrease the survival rate of the juvenile age class.

9. Under the *Model* menu select *Stage Matrix*. Change the survival rate of the juvenile age class from 0.28 to 0.18. Click OK to exit this window.
10. Now you are going to run the simulation. Under the *Simulation* menu, select *Run*, or press *Ctrl-R* on your keyboard. At the bottom right corner of this window you will see a message when the simulation is complete. At this point you may close the window by clicking on the X in the upper right corner of the window.
11. Under the *Results* menu, select *Trajectory Summary*. Here you can view the population growth curve. The time scale, like the age classes, is in 1 year time steps. Print the plot of the *Trajectory Summary* by clicking on the printer icon.
12. You can view the numbers corresponding to the plot by clicking the *Show Numbers* button in the upper left corner of the *Trajectory Summary* window. Record the average population size at time steps 0, 10, 20, 30, 40, and 50 in the appropriate column of Table 10.6.
13. You can return to the plot by clicking the *Show Plot* button in the upper left corner of the *Trajectory Summary* window. Exit the *Trajectory Summary* window by clicking on the X in the upper right corner of the window.
14. Now suppose that you underestimated the carrying capacity of the population. Under the *Model* menu select *Stage Matrix*. Change the survival rate of the juvenile age class from 0.18 to 0.28. Click OK to exit this window.
15. Under the *Model* menu select *Density Dependence*. Change the *Carrying Capacity* from 60 to 100. Click OK to exit this window.
16. Repeat steps 10 through 13.
17. Make a line graph comparing the average abundance at time steps 0, 10, 20, 30, 40, and 50 for simulations with a carrying capacity of 60 to those with a carrying capacity of 100.
18. For the simulation with a carrying capacity of 100, record the minimum and maximum values at time steps 0, 10, 20, 30, 40, and 50 in Table 10.7. Calculate the difference at each time step. Enter the values in the appropriate column of Table 10.7.

Table 10.7. Difference between minimum and maximum at each time step.

Time Step	Minimum	Maximum	Difference
0			
10			
20			
30			
40			
50			

19. Make a line graph of the difference between the lowest and highest population estimate versus time.

Questions

1. In general, does the difference between the minimum and maximum population size predicted increase or decrease over time? Why?

2. What does this pattern indicate about the ability to predict the viability of a population far into the future?

3. If you cannot predict the future size of a population with much accuracy, why should you use population viability analysis?

4. What is the effect if you incorrectly estimated juvenile survivorship at 0.28, assuming it was actually 0.18?

5. What effect did increasing the carrying capacity have on the average population abundance?

6. Make a list of factors that we need to be concerned with when we formulate a reintroduction program.

> *Your lab report should include the following:*
>
> 1. A population trajectory graph, an abundance table (Table 10.1), and answers to question 1 and 2 for Part 1 of Exercise A
> 2. An abundance table (Table 10.3), a population trajectory graph, and answers to questions 1 through 5 for Part 2 of Exercise A
> 3. A standard deviation table (Table 10.4), an abundance table (Table 10.5), 1 line graph, and answers to questions 1 through 3 for Part 3 of Exercise A
> 4. An abundance table (Table 10.6), 2 line graphs, a min-max table (Table 10.7), and answers to questions 1 through 6 for Exercise B

References

Griffith, B., J. M. Scott, J. W. Carpenter, and C. Reed. 1989. Translocation as a species conservation tool: Status and strategy. *Science*, 245: 477–480.

Salz, D. and D. I. Rubenstein. 1995. Population dynamics of a reintroduced Asiatic wild ass (*Equus hemionus*) herd. *Ecological Applications*, 5(2): 327–335.

Seal, Ulysses S., (ed.) 1989. *Conservation Biology and the Black-Footed Ferret*. Yale University Press, New Haven, CT.

Laboratory 11
Park Size and Species Diversity: Lessons from Islands

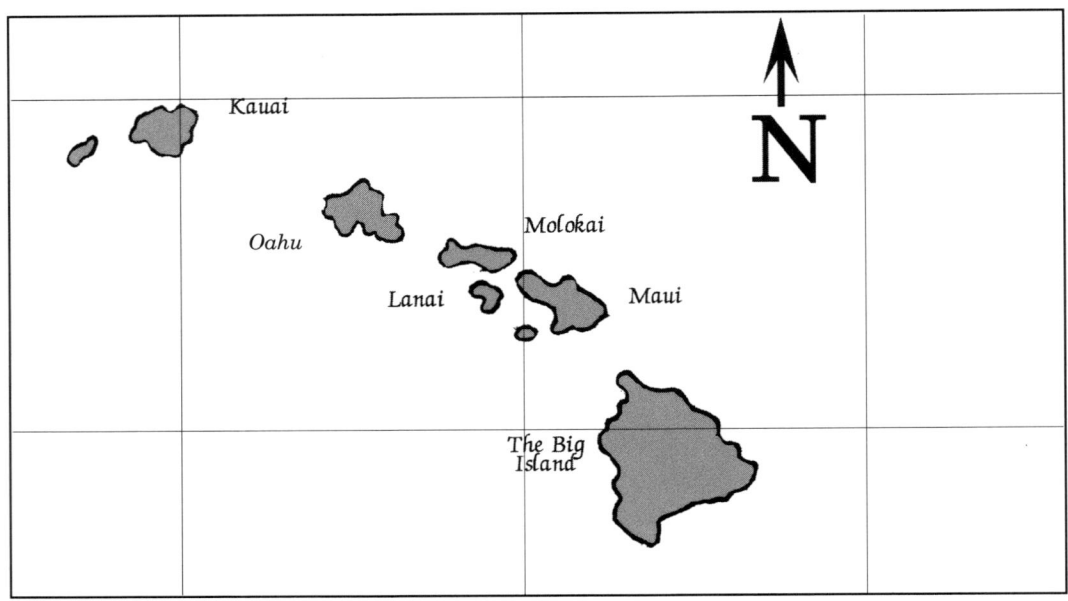

Island biogeography is an important area of study that can provide guidelines for conservation. Many of the natural areas left on earth are "islands" in the middle of unsuitable or developed habitat. As these protected areas decrease in size, the number of species likely to persist declines. In this lab you will learn about the predictions of island biogeographic theory and its implications for the protection of biodiversity.

Introduction

Ecologists Robert MacArthur and E. O. Wilson introduced the theory of island biogeography in the 1960s. Developed originally to predict community patterns on islands, now this theory is used to predict and evaluate species occurrences in fragmented habitats. MacArthur and Wilson's theory attempts to explain the patterns of species composition seen on islands of different sizes and different distances from the nearest mainland. They published a short book titled *The Theory of Island Biogeography* in 1967, which introduced the idea of species-area relationships.

The species-area relationship illustrates how the size of an island is directly proportional to the number of species on that island. Therefore, the larger the island, the more species are found on it. They theorized that this is because large islands have a much lower species extinction rate than small islands. MacArthur and Wilson also noticed that the farther an island is from the mainland, the longer it takes for new species to colonize. The balance between colonization and extinction results in different numbers of species that can be supported by different types of islands. A large island near the mainland will have high colonization and low extinction, thereby supporting many more species than a small island that is far away from the mainland. Figure 11.1 illustrates the relationship between the rates of colonization and extinction of islands.

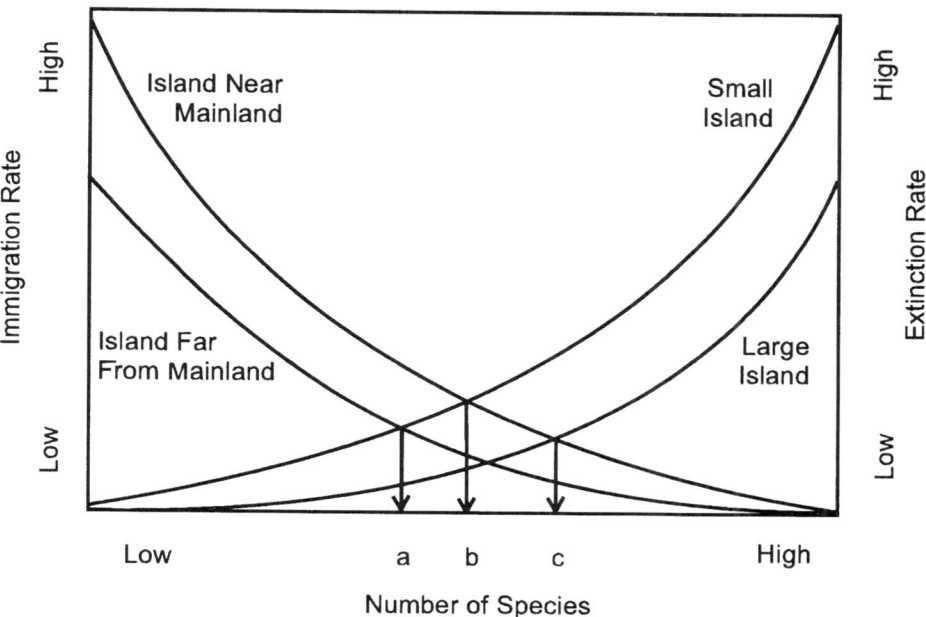

Figure 11.1. The island biogeography model describes the relationship between the rates of immigration and extinction on islands (after MacArthur and Wilson, 1967). Shown are the number of species likely to be supported for a (a) small island far from the mainland, (b) small island near the mainland, and (c) large island near the mainland.

Simberloff and E. O. Wilson tested these ideas by looking at species richness and colonization of mangrove islands in Florida (Simberloff and Wilson, 1970). They removed all the insects from mangrove islands and observed the rate of colonization and the final number of species on each island. They found that islands that were larger or closer to the mainland indeed contained more species than islands that were smaller and farther away from the mainland.

Area and the Number of Species

Islands generally have fewer species and higher extinction rates than do continental environments. The theory of island biogeography can explain the species differences between different islands and those between islands and the mainland. Ecologists realized that the theory of island biogeography also may be applied to discontinuous patches of habitat on continents. Thus, the species-area relationship can be used to predict the number of species that can be supported by isolated habitat patches such as forest fragments.

If protected land is surrounded by habitat that is disturbed either by farming, deforestation, development, or pollution, the intact habitat is an island in a sea of hostile land. A small-scale example of habitat patches, or habitat islands, is parks within an urban area. City parks are very much islands of habitat surrounded by an inhospitable environment. City parks also come in varying sizes and can be different distances away from the nearest forest or undeveloped land. Other examples of habitat "islands" include forest fragments in the middle of agricultural land, fresh water lakes that are separated by dry land, and small patches of specific habitat types.

MacArthur and Wilson noticed that although the number of species increases with habitat area, the relationship is not a straight line. As a rule of thumb, for every tenfold increase in magnitude (e.g., from 10 to 100), the number of species doubles. The size of an area affects each environment and group of species a little differently. It is possible to use a formula to take into account the differences between environments. A generalized species-area relationship is as follows:

$$S = K \times A^z$$

(Number of species) = (species scale) × (area of the island)^z

In this equation, the species scale (K) indicates how many species are found per area in a given environment. The variable z represents how the area affects the number of species found. The larger the value of z, the more species that are added for a given amount of area. Figure 11.2 illustrates an example of this type of relationship for the number of bird genera (singular: genus) found on different size islands in the South Pacific.

Notice that the number of genera does not increase linearly, or in a straight line, as the area of the island increases. If one takes the logarithm of the area, though, this relationship will become a straight line, making the relationship easier to interpret. When the relationship between two variables is curved and not linear, a logarithm transformation of the variable on the x-axis is likely to produce a linear graph. Figure 11.3 presents the same data as in Figure 11.2 except that the logarithm of the island area is now plotted on the x-axis.

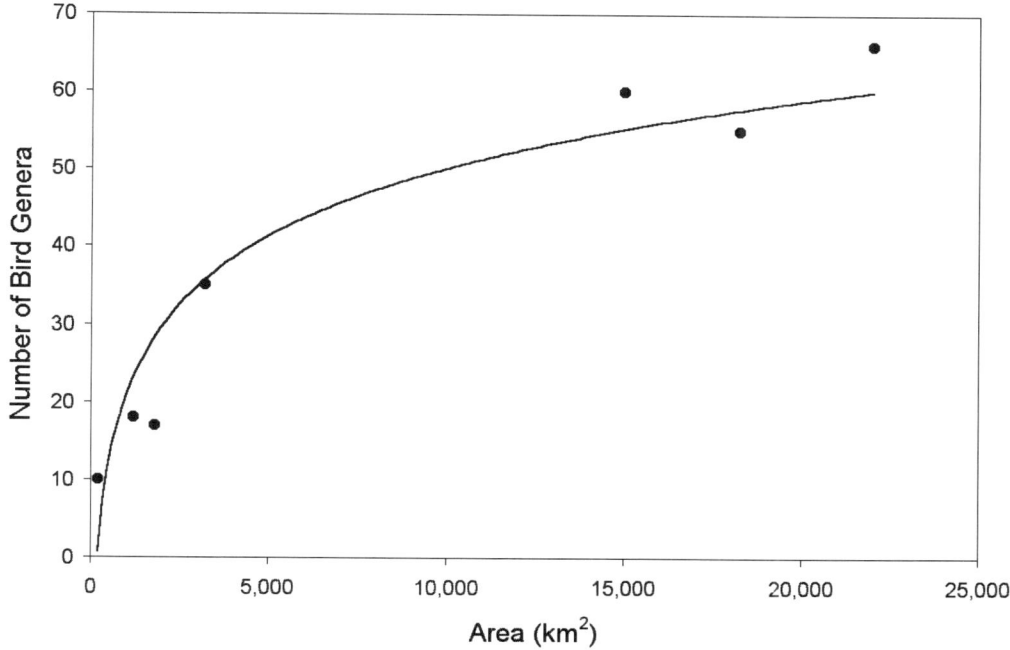

Figure 11.2. The bird diversity versus island size in the South Pacific (based on data from Cox and Moore 1993).

Island Biogeography and Reserve Design

During the 1970s, a debate developed in ecology and conservation biology over how to apply the theory of island biogeography to conservation and reserve design. On one hand, the larger a reserve is, the more species it holds and therefore protects. In 1982, a survey by the IUCN of the world's national protected areas and parks revealed that one half of the protected areas are less than 100 km^2 (38.6 miles^2) and 98% of the parks are less than 10,000 km^2. (3861miles^2). On the other hand, two small reserves may be able to preserve more species than one large reserve, especially if each small area protects more types of habitats than a single large one. The debate was termed SLOSS, meaning *single large or several small* reserves. Of course, the best option is to have many large reserves, but that is not practical in all situations.

Consider a large unexploited forest. As conservation advocates, we may want to designate some of this forest as a reserve. If we are restricted to preserving a limited amount of forest, and the rest will be logged and developed during the next several decades, how do we determine what areas and in what pattern the reserve or reserves should be created?

One viewpoint is that large preserves are needed to protect species that live at low densities. Several small reserves will not be able to support species that require large tracts of land. In the United States some of the species for which this argument has been made are grizzly bears, wolves, spotted owls, and the Florida panther. Many mammals and birds cover very large areas regularly foraging for food and looking for mates. Often these home range areas are too large to fit within the boundaries of small parks. For example, the

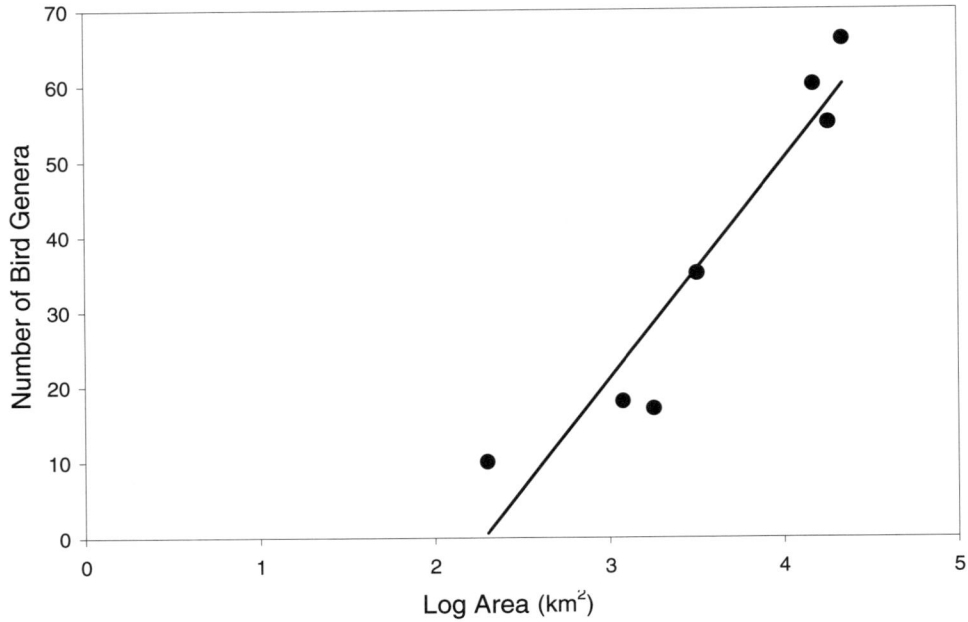

Figure 11.3. A logarithm transformation of the bird diversity versus island size found in Figure 11.2.

cougar in southern California has an average home ranges size of 450 km^2 (173.7 miles^2) for males and 155 km^2 (59.8 miles^2) for females. Based on IUCN's 1982 estimates, more than half of the world's parks would not be large enough to support a single female cougar (assuming the habitat was suitable).

The species-area relationship described earlier predicts that a large reserve will maintain more species than a small one. Even if there are many small reserves, as long as they are made up of the same type of habitat, it is likely that they are all home to the same species, and the total number of species will be less than a single large reserve. However, if each small reserve is composed of different habitat types, and therefore contains different species, the total species number of several small reserves may be much higher than the total of the large one.

Consider the following situation: There is a natural area that is composed of several habitat types. The majority of the natural area is made up of hardwood forest and is home to deer, owls, salamanders, squirrels, and many other forest species. Additionally there is a small wetland area that is home to several rare frogs, a few songbirds that only nest in wetland areas, some freshwater fish, a few migratory ducks, beavers, and muskrats. Another habitat type found in the natural area is a small, open grassland that supports butterflies, burrowing rodents like ground squirrels and voles, burrowing owls, tortoises, and quail. Because the forest makes up the majority of the land, if we decide to protect a large continuous area that contains 50% of the total area, chances are the protected land will contain only forest habitat. But with careful planning, the allotted area can subdivided into 3 smaller reserves so that the different habitat types, and the diversity of species within each, are all protected.

Another advantage of having many small reserves instead of a single large one is that separate reserves may not all be affected by harmful events. A fire may destroy one small reserve but will not be able to spread to others. Not only does this protect the critical habitat of the remaining reserves, but it generates more habitat variability (and therefore more species variability) between reserves. Diseases that affect a population in one park may not infect populations in separate parks. Invasive species, such as rats or weeds, likewise may only affect one or few of the small reserves. If we only protect one large area, any adverse condition may affect all of the protected area, not just portions of it.

Fragmentation and Edge Effects

When we divide a habitat into fragments, we also need to consider how fragmentation affects habitat quality. **Edge effects** are differences in habitat structure and species composition near the boundaries of a habitat patch. A common example of edge effects is the weeds and shrubs that grow at the border between a field or lawn and a nearby forest. The species that are found at this edge are different from species inside the forest or within the field. Animal species are often more vulnerable to predation and nest parasitism at forest edges (less cover to hide from predators). The smaller the habitat patch is, the more important edge effects become. Small patches are more effected by edge effects because they have a larger edge to interior ratio.

Songbirds often have lower nesting success near the edge of forests. Nest predators such as the blue jay, eastern chipmunk, and raccoon, as well as brood parasites such as the brown-headed cowbird, occur in higher numbers near the forest edges. Figure 11.4 illustrates this correlation between the increased nesting success and decreased levels of predation for songbirds in Michigan.

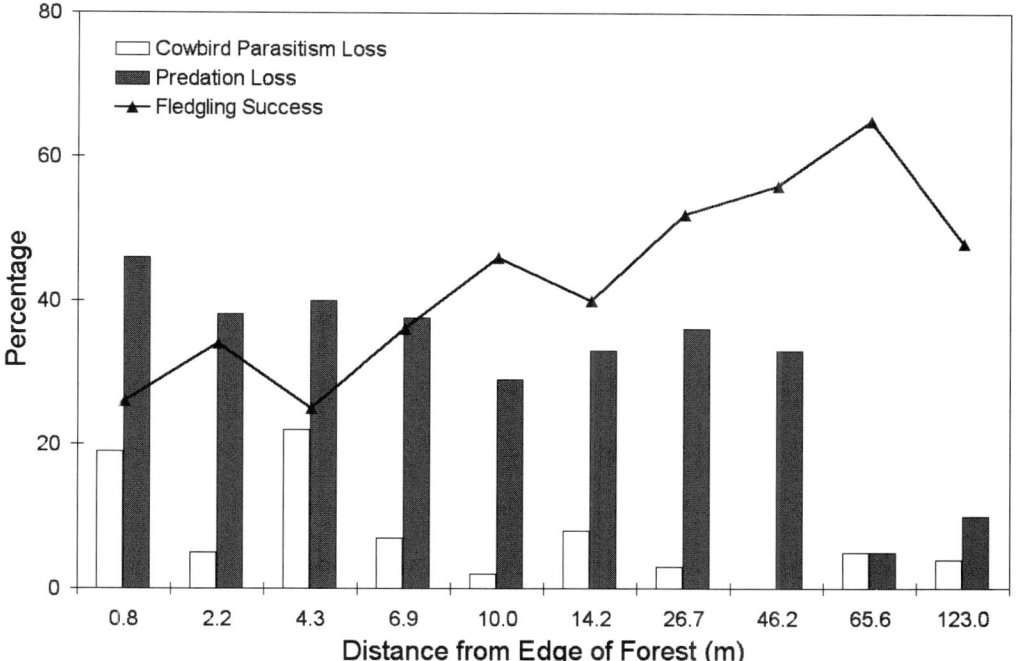

Figure 11.4. Nesting success of Michigan songbirds as the distance from the forest edge increases compared to the level of predation (after Gates and Gysel 1978).

In the 1970s, Thomas Lovejoy began a long-term project to look at reserve design called the Minimum Critical Ecosystems Project (Lovejoy et al. 1986). He convinced loggers in the Amazon basin to leave forest patches of different sizes untouched so that he could determine the effect of patch size on the abundance and diversity of species within them. Over time, the smallest forest fragments lost many of the species that were originally found there. Strong edge effects have made patches unsuitable for many sensitive species. Entire populations of birds and monkeys have died out in many patches. This project has shown that in several decades forest fragments change dramatically. The long-term effect of small patch size is very poorly understood because we do not have data on changes in species composition over time.

Applications in Conservation Planning

The question still remains of how to use species-area relationships and island biogeography in conservation planning. Although island biogeography allows us to make predictions about the stability of species richness once habitat is fragmented, it does not inform us which species will be protected and which species will drop out of the community. Additionally, the specific characteristics of an environment, and not just the area, often will provide more information about the stability of populations and species. If we are trying to protect species that occur only in subhabitats, including all of that habitat type may be more important to the species than the number of protected hectares.

The stability of a population also may be largely affected by the degree of habitat connectivity. If individuals easily can leave one habitat fragment and find another, each fragment may be small enough to support only a few individuals, but many fragments together can support a substantial population. You will learn more about metapopulations, or separate but connected populations, in a later lab.

Exercise A: The species-area relationship

Background

In this exercise, you will consider National Parks and other protected areas in northern California and Florida to look for species-area relationships in actual parks. You will focus on mammal communities in northern California and amphibian communities in Florida. The data for this exercise is from the Information Center for the Environment (ICE) database available on the World Wide Web (www.ice.ucdavis.edu). Your instructor may decide to use data from that source in addition to, or in place of, the information in this exercise.

Protected areas in California are generally composed of pine or other coniferous forests, with some grassland and other habitats. The parks considered in this exercise range from those located near the coast to those that are high in the Sierra Mountain Range. The number of mammal species in a few of the protected areas is shown in Table 11.1. In contrast, protected areas in Florida, Georgia, and South Carolina are mostly composed of cypress, swamp, and coastal vegetation. The number of amphibian species in some southern protected areas is shown in Table 11.2.

As you investigate species-area relationships for mammals in California and amphibians in Florida, keep in mind that different groups of organisms, namely, mammals and amphibians, differ in their basic habitat requirements. Amphibians, in general, require very specific conditions to survive. Compared to most mammal species, amphibians are not good at dispersing between reserves, and they seem to have a more difficult time maintaining populations in small reserves. Also, amphibians are very sensitive to toxins and pollutants in their environment.

Table 11.1. Number of mammal species in California protected areas of various sizes.

Park Name	Area (hectares)	Log Area	Mammal Species
Muir Woods	224		24
Pinnacles National Monument	740		49
Whisky-Shasta Trinity National Recreation Area	17,200		65
Lassen National Park	43,048		66
Yosemite National Park	308,068		93
Sequoia-King's Canyon National Park	349,310		90

Table 11.2. Number of amphibian species in southern protected areas of various sizes.

Park Name	Area (hectares)	Log Area	Amphibian Species
Ocmulgee National Monument	284		6
Congaree Swamp National Monument	8984		13
Cumberland Island National Seashore	14,737		14
Gulf Island National Seashore	54,879		11
Big Cypress National Preserve	289,761		21

Exercises

1. Using the information on California parks from Table 11.1, create a graph in which you plot reserve size against the number of species found in each park. You may

generate this graph by using a graphing program or by hand. In your graph, the *x*-axis should be the park size and the *y*-axis should be the species number.
2. Using a calculator, fill in the appropriate column of Table 11.1 with the log of the reserve area.
3. Create a second graph of park area versus species abundance, but this time use the logarithm of the park size as the *x*-axis. Plot the species number on the *y*-axis.
4. Using the information on Florida parks from Table 11.2, create a graph in which you plot reserve size against the number of species found in each park. You may generate this graph by using a graphing program or by hand. In your graph, the *x*-axis should be the park size and the *y*-axis should be the species number.
5. Using a calculator fill in the appropriate column of Table 11.2 with the logarithm of the reserve area.
6. Repeat step 3 using the information from Table 11.2.

Questions

1. What is the relationship between the size of the park and the species number for the first graph you made? If you drew a line to show how park size relates to species number, what general shape would this line have (a straight line? a curve?)?

2. When you use the log of the park area, is the relationship between species number and park area easier to interpret? What would the line that shows the relationship of the log of park size to the species number look like?

3. Compare the graphs of mammals in California to those of amphibians in Florida. Are they similar? What is a general pattern to describe both sets of graphs?

4. How do the number of amphibian species differ from the number of mammal species in parks of approximately the same size? What might account for this difference?

5. What would be the value of understanding the relationship between area and the number of species in a given region?

Exercise B: Predicting species abundance

Background

In this exercise, you will predict the number of species that will occur in reserves of different sizes based on the formula for the species-area relationship. Then, you will explore the SLOSS debate by making different assumptions about the habitats that make up these parks.

Recall the species-area relationship described earlier. The species scale, K, is the number of species found per area in a specific environment. You are going to be looking at reptiles in a hypothetical scrub community; the K value for the animals will be 6. Researchers have found that most islands or isolated parks have a z value between 0.24 and 0.35, whereas most continental areas or large tracts of land have a z value between 0.15 and 0.22. For the following exercises, the scrub environment has a z value of 0.2.

$$S = K \times A^z$$

(Number of species) = (species scale) × (area of the island)^z

Exercises

1. Using the species-area formula above, fill in Table 11.3 with the expected number of reptile species found in a reserve of the following sizes.

Table 11.3. The number of reptile species found in reserves of different sizes.

Reserve Size (hectares)	Log Reserve Size	Number of Reptile Species
200		
500		
1000		
5,000		
10,000		
50,000		
100,000		

2. Using the information from Table 11.3, create a graph in which you plot reserve size against the number of reptile species. You may generate this graph by using a graphing program or by hand. In your graph, the x-axis should be the reserve size and the y-axis should be the species number.
3. Using a calculator, calculate the logarithm of the reserve size and fill in Table 11.3. Create a second graph by plotting the logarithm of the reserve size against the expected number of species.
4. Now you will consider how the value of z might affect the estimate of species number. Using the formula for species-area relationships, fill in Table 11.4, using different values of z.

Table 11.4. The number of reptile species found in reserves of different sizes, using different values of the exponent z.

Reserve Size (hectares)	Number of Reptile Species		
	$z = 0.15$	$z = 0.20$	$z = 0.22$
1000			
5,000			
10,000			
50,000			
100,000			

Questions

1. Are you likely to find exactly the expected number of species in any of the reserves? Explain.

2. What are some reasons you may not find the expected number of species in a park?

3. If you harvest half of the plants in a 5000 hectare patch of habitat, how many reptile species would be lost (based on your table)? What percentage of the total would be lost?

4. If you log 90% of the 5000 hectare patch of habitat, how many (and what percentage) of reptile species would be lost?

5. Based on the species-area relationship, would it be better to buy two 500 hectare plots of land or a single 1000 hectare plot, in terms of the number of species that would be protected? On what does your answer depend?

6. How does the number of species change if you change the value of z?

7. When might a single large reserve be a better option than several small reserves? Think in terms of the ecology and habitat types in different situations.

8. Under what circumstances might several small reserves be better than a single large reserve?

Your lab report should include the following:

1. The completed Tables 11.1 and 11.2, 4 scatter diagrams, and answers to questions 1 through 5 for Exercise A
2. The completed Tables 11.3 and 11.4, 2 scatter diagrams and answers to questions 1 through 8 for Exercise B

References

Cox, C., and P. Moore. 1993. *Biogeography: An Ecological and Evolutionary Approach.* Blackwell Scientific, Boston, MA.

Gates, J. E., and L. W. Gysel. 1978. Avian nest dispersion and fledgling success in field-forest ecotones. *Ecology*, 59: 871–883.

Lovejoy, T. E., R. O. Bierregaard, Jr., A. B. Rylands, J. R. Malcolm, C. E. Quintela, L. H. Harper, K. S. Brown, Jr., A. H. Powell, G. V. N. Powell, H. O. R. Schubart, and M. B. Hays. 1986. Edge and other effects of isolation on Amazon forest fragments. In Soulé, M. E. (ed.), *Conservation Biology: The Science of Scarcity and Diversity*, pp.257–285. Sinauer Associates, Sunderland, MA.

MacArthur, R. H., and E. O. Wilson. 1967. *The Theory of Island Biogeography.* Princeton University Press, Princeton, NJ.

Simberloff, D. S., and E. O. Wilson. 1970. Experimental zoogeography of islands: A two-year record of colonization. *Ecology*, 51: 934–937.

Laboratory 12
Rescuing the Spotted Owl: Conserving Species in Multiple Populations

As forests become more and more fragmented due to logging and development, populations that once inhabited large continuous areas also become fragmented. The dynamics of these multiple populations that interact through limited dispersal differ from the single isolated populations you have been dealing with in previous laboratories. Here, you will explore patterns of extinction and recolonization of fragmented and patchy habitats using your knowledge of single populations and applying new concepts dealing with dispersal between multiple populations.

Introduction

Until now, we have focused on the dynamics of single populations. In the single population examples of previous laboratories, we assumed that there was no dispersal (immigration or emigration) between populations. This approach is very useful when one is concerned with the problems of isolated populations, such as managing grizzly bears in a single nature reserve or the white rhino populations confined by fences in Zimbabwe. The single population approach is also useful for species that live in large uniform habitats, such as the blue whale or the bluefin tuna.

Many species, however, exchange some individuals between populations through dispersal. For these species, the single population models that we have been using will not be sufficient for understanding how the populations change over time. In these cases we must adopt a multiple population approach in which the characteristics of each population and the effect they have on each other are considered.

Populations that exchange individuals through immigration and emigration form a single unit called a **metapopulation**. Each population that makes up a metapopulation is referred to as a subpopulation, local population, or simply as a population. Understanding the interactions between subpopulations is important when designing reserves and for planning the least intrusive patterns for natural resource extraction such as forest cutting. Dispersal will affect the extinction rates, the degree of inbreeding, and the rate of recolonization of extinct subpopulations.

As individuals migrate into a subpopulation, they not only affect its size, but they introduce genes that previously may not have been present, thereby increasing the population's genetic diversity. In a quickly growing population, surplus individuals may disperse to another area and prevent overcrowding. Dispersal may also allow a recently extinct population to become reestablished through immigrating individuals from an extant (surviving) population.

How Do Metapopulations Form?

Metapopulations can result from naturally patchy habitats or from human disturbance resulting in habitat fragmentation. For example, patchy environments, such as a series of ponds, a mountain range full of peaks and valleys, or patches of forest surrounded by human development, often contain species living in metapopulations. Environments such as these are isolated enough to prevent individuals from acting as a single population where all individuals may freely interact and interbreed. However, the environments surrounding the habitat patches often allow a limited exchange of individuals to occur between patches.

A metapopulation cannot be treated as a single population because the interacting subpopulations each have their own set of fecundity and survival values. The different patches may have slightly different communities of species that affect food availability, competition with other species, or predation. The fecundity will vary because the habitats often differ in food quality and abundance, or resources needed for reproduction, such as nest sites. Predation rates or other sources of mortality also may vary between sites.

Abiotic factors, such as soil composition, water quality, inorganic nutrients, and sunlight, may differ even between areas that are close together. Subpopulations of a plant species, for example, may be located in areas differing in soil nitrogen availability (a nutrient that may limit the growth of a plant). This may result in some subpopulations producing more seedlings than those in other areas. Also, certain random events may not affect all subpopulations in the same way. For example, plant populations living in areas of different moisture levels will be affected differently when exposed to a wildfire. In the same way, a catastrophic event such as a tornado is unlikely to damage every subpopulation equally.

There are several factors that will affect the chance of a metapopulation becoming extinct. These include factors that increase the chance of extinction in each local population, how closely the dynamics of the subpopulations reflect each other (their correlation), and factors that affect dispersal patterns between the populations. We have previously discussed the factors that affect extinction rates in a single population; therefore, we now will focus on the dispersal rates and correlation among subpopulations of a metapopulation.

Correlation among Populations

If two local populations have correlated dynamics, their fluctuations in population size will be synchronous (they will fluctuate together). In other words, if one of these populations goes extinct, there is a good chance that the other one will do the same. In this case, the risk of extinction of each individual population is similar to the risk of extinction of the entire metapopulation. This may be the case for 2 subpopulations that are very close together geographically. They will have very similar habitats and will constantly be exposed to the same environmental changes. A good year in one population will probably be a good year in the other population.

On the other hand, if 2 subpopulations experience very different environmental fluctuations, as is often true of populations that are far apart, the extinction probabilities are not correlated. To estimate the risk of extinction of the metapopulation as a whole, one multiplies together the individual probabilities of extinction of each subpopulation. Even if the 2 subpopulations have the same individual risk of extinction, the metapopulation will have a lower risk of extinction. In general, the metapopulation will always have a lower risk of extinction than any of its component subpopulations, because the probability of all the populations becoming extinct at once is usually very low.

An analogy can be made with rolling a pair of dice. Because there are 6 numbers on a die, each has a 1 out of 6 chance of showing the number 1. The probability of rolling two 1s on the same roll, however, is not 1 out of 6. To determine this probability, you must multiply the probabilities of rolling a 1 on each die. This means that there is a $(1 \div 6) \times (1 \div 6) = 0.028$ chance of rolling snake eyes. Two populations that have independent chances of extinction, because the environmental fluctuations affecting them are not correlated, can be treated like the dice in this example. The risk of extinction for the metapopulation as a whole can be determined by multiplying the risk of extinction for each subpopulation.

As population fluctuations become more synchronized, the risk of extinction for the metapopulation becomes greater because subpopulations are likely to become extinct at the same time. If two local populations are in close proximity to one another, it is likely that a disturbance that affects one will affect the other similarly. Likewise, if a population is spread out over a number of populations, the extinction risk for the metapopulation becomes lower because they are not all likely to become extinct at the same time.

Remember that the technique of multiplying probabilities can only be applied if the probabilities of extinction of subpopulations are independent of each other. The farther apart the populations are, the more likely it is that the extinction probabilities are not correlated. For example, Thomas (1991) found that silver-studded butterfly populations that were geographically close tended to fluctuate in synchrony whereas populations that were further apart fluctuated independently of one another.

Considering this, one might expect that a metapopulation made up of widely dispersed subpopulations would have a low extinction risk. However, the geographical configuration of populations also affects the dispersal rates among them. Very low dispersal between subpopulations may result in increased inbreeding within these local populations (causing lower genetic diversity) and also decrease the chance of a recolonization event. Both of these effects will increase the chance of extinction for a subpopulation. In addition to distance effects, factors that affect dispersal such as habitat barriers, abundance of individuals, sex and age compositions of populations, and stochastic events also should be considered. If a local population becomes extinct, individuals from an extant (existing) population may recolonize it. Therefore, dispersal among local populations that leads to successful recolonization usually decreases extinction risks of the species.

In a previous lab you experimented with translocating rhinos into a declining population. The addition of animals stabilized the population. In the same way, natural dispersal from productive patches can act to prevent a local population from becoming extinct. Dispersal can work to prevent extinction on both the local and the metapopulation level.

Without the influx of individuals from productive habitats, some subpopulations, located in low quality habitats, might eventually become extinct. The subpopulations that must be maintained through immigration are called sink populations. These areas often have high mortality rates and/or low birth rates. Productive habitats often produce source populations. These have high or stable growth rates even though emigration exceeds immigration. Intuitively you might think that a sink population is negative for a metapopulation as a whole, but this is not necessarily true. If individuals are being produced in excess of the carrying capacity of one area, they may disperse to a sink population allowing the metapopulation to hold more individuals. In some species, for example, pre-breeding subadults may leave a source population and live in a sink habitat until they are able to reproduce. At which point they may return to the more productive habitat of the source population. If the sink habitat were removed, the subadults would have to remain in the source population and compete with the breeding adults for resources. Because of density dependent factors, the metapopulation will hold fewer individuals.

Patches that are close to one another have high dispersal and recolonization rates that lower the risk of extinction. Patches that are far apart are likely to have uncorrelated probabilities of extinction that also lower the risk of extinction. Conservation biologists must, therefore, consider the trade-off between these distance effects when making management decisions for a species in a metapopulation.

Exercise A: The southern California spotted owl

Background

The spotted owl of California is an endangered species. Development and forest clearing by humans have fragmented its preferred habitat. The remaining suitable habitat patches are of different sizes and distances from one another. The model in this exercise is based on one of the models used by LaHaye et al. (1994) to explore the effects of spatial factors in this metapopulation.

Exercises

Part 1: The initial conditions

1. Open the program RAMAS EcoLab (double-click on its icon with your mouse).
2. Click on Multiple Populations.
3. Under the *File* menu, select *Open* and choose *Owl.mp*. This file contains a metapopulation model of the California spotted owl. A map of the distribution of owl populations in California will appear. The circles represent the subpopulations of owls. The larger the circle is, the higher the carrying capacity of the reserve. The lines signify the dispersal routes used by the owls. If you like, you may make the map larger by clicking on the "maximization" button in the upper right corner of the window, to the left of the X.
4. Under the *Model* menu select *General Information*. Set the *Duration* to 20. By doing this, you have set the program to simulate 20 time steps of reproduction. In this exercise, each time step is 1 year long.
5. Also in the *General Information* window, set the number of *Replications* to 100. Make sure there is a checkmark in the box for *Use demographic stochasticity*. If there is no checkmark, click the box to make one appear. Each of the 100 replications will differ slightly because of the demographic variation. Click OK to exit this window.
6. Under the *Model* menu select *Population*. By clicking on the name of individual populations you can view the characteristics of each population. Notice that all of the populations experience exponential growth (lack of density-dependent growth) with a *Growth rate* of 0.8270 and a *Survival rate* of 0.750. Click *Cancel* to exit this window.
7. From the *Model* menu select *Dispersal and Correlation*. Notice that the *Maximum dispersal rate* allowed in the model is 0.05, or 5%, of the population. This means that at every time step a maximum of 5% of each population may emigrate to another population, using the dispersal routes indicated on the main map.
8. Also in the *Dispersal and Correlation* window, notice that the *Average dispersal distance* is set at forty with a *Maximum dispersal distance* of 100. Click *View Function* next to the *Maximum dispersal distance*. Here you can see the effect distance has on the amount of dispersal. From this you can infer that populations separated by 50km (31 miles) exchange less than 1% of their inhabitants. The maximum dispersal rate (5%) is only achieved by very close populations. Close this window by clicking on the X in the upper right corner.
9. Also in the *Dispersal and Correlation* window, notice that the *Correlation* among distant populations is set at 1.0. This indicates that there is complete correlation of environmental fluctuations between populations. Click *Cancel* to exit this window.
10. Now you are going to run the simulation; under the *Simulation* menu select *Run*, or press *Ctrl-R* on your keyboard. As the simulation runs, the colors of the circles on the map will change to reflect the changing sizes of the populations. The model will run much faster if you view the text rather than the map. To do so, click on the leftmost icon in the simulation window as the simulation is running. At the bottom right corner of this window you will see a message when the simulation is complete. At this point you may close the window by clicking on the X in the upper right corner of the window.

11. Under the *Results* menu, select *Trajectory Summary*. Here you can view the population growth curve of the metapopulation as a whole (referred to as *Population 0*). The time scale is in 1 year time steps.
12. You can view the numbers corresponding to the plot by clicking the *Show Numbers* button in the upper left corner of the *Trajectory Summary* window. You can return to the plot by clicking the *Show Plot* button also in the upper left corner of the *Trajectory Summary* window.
13. You can also view a plot and table of population growth for individual subpopulations (*Populations 1–22*). To do this, click the up or down arrow in the box next to the word *Population* at the top of the *Trajectory Summary* window.
14. In the table below, record the *Average Abundance* of the metapopulation (*Population 0*) at time steps 1, 5, 10, 15, and 20. Also record the *Average Abundance* of *Populations 3, 6, 18,* and *19* at the same time steps. Exit the *Trajectory Summary* by clicking on the X in the upper right corner of the window.
15. Estimate the percentage change from the initial abundance for the metapopulation and *Populations* 3, 6, 18, and 19. Calculate this change using the formula:

[(Final abundance − initial abundance) / initial abundance] × 100

Enter the values in Table 12.1.

Table 12.1. Average abundance for some spotted owl populations.

	Average Abundance for Each Population				
Time	All Pops	Pop 3	Pop 6	Pop 18	Pop 19
0					
5					
10					
15					
20					
% Change					

Questions

1. What is a metapopulation?

2. Why should the modeling of metapopulations be approached differently from that of single populations?

3. Name three factors that affect the extinction rate of metapopulations.

4. What is expected to happen to the spotted owl metapopulation if it is allowed to exist with the survival, fecundity, and dispersal values used in this model? Be specific.

5. In this metapopulation model, do all populations show the same trend in abundance over time? Explain why this is so.

Part 2: The effects of dispersal

Now you will investigate the potential effects of decreasing dispersal among the populations. Suppose that development (houses, roads, factories, etc.) occurred among the suitable habitat patches, making it more difficult for the spotted owl to disperse from one population to another. What effect do you think this would have on the final abundance of the metapopulation?

1. From the *Model* menu select *Dispersal and Correlation*. Reduce the amount of dispersal in this metapopulation by changing the *Average dispersal distance* to from 40 to 5, and changing the *Maximum dispersal distance* from 100 to 25. Click OK to exit this window.
2. Now you are going to run the simulation; under the *Simulation* menu select *Run*, or press *Ctrl-R* on your keyboard. The model will run much faster if you view the text rather than the map. To do so, click on the leftmost icon in the simulation window as the simulation is running. At the bottom right corner of this window you will see a message when the simulation is complete. At this point you may close the window by clicking on the X in the upper right corner of the window.
3. Under the *Results* menu, select *Trajectory Summary*. Here you can view the population growth curve of the metapopulation as a whole (referred to as *Population 0*). The time scale is in one-year time steps.
4. You can view the numbers corresponding to the plot by clicking the *Show Numbers* button in the upper left corner of the *Trajectory Summary* window. You can return to the plot by clicking the *Show Plot* button, also in the upper left corner of the *Trajectory Summary* window.

5. In the table below, record the *Average Abundance* of the metapopulation (*Population 0*) at time steps 0, 5, 10, 15, and 20. Also record the *Average Abundance* of *Populations* 3, 6, 18, and 19 at the same time steps. Exit the *Trajectory Summary* by clicking on the X in the upper right corner of the window.
6. Estimate the percentage change from the initial abundance for the metapopulation and *Populations* 3, 6, 18, and 19. Calculate this change using the formula:

[(Final abundance − initial abundance) / initial abundance] × 100

Enter the values in Table 12.2.

Table 12.2. The average abundance for some spotted owl populations with dispersal.

	Average Abundance for Each Population				
Time	All Pops	Pop 3	Pop 6	Pop 18	Pop 19
0					
5					
10					
15					
20					
% Change					

Questions

1. Is the final abundance for the metapopulation more, less, or the same when the maximum dispersal distance is decreased?

2. Do all of the populations respond to the decreased level of dispersal the same way? Is the effect greater or smaller or the same in small compared to large populations? Why?

Exercise B: Designing reserves for the spotted owl

Background

The single largest factor affecting the metapopulation of spotted owls in California appears to be the decline in habitat quality in recent years. In this activity you will assume that we can improve the habitat but that it will cost a lot of money. The cost for improving the habitat occupied by each subpopulation is $1000 multiplied by the carrying capacity of area. By improving an area, it is possible to increase the growth rate of the subpopulation to 1.01. You are asked to determine the best way to spend $500,000 to conserve the spotted owl in California.

Exercises

Restoration alternatives

You are now going to simulate the restoration of different subsets of populations to investigate how populations with different characteristics affect the metapopulation as a whole. You will first restore a random subset of populations, then only the smallest or largest populations, and finally only those populations that are very close to or very far from one another.

1. Under the *File* menu, select *Open* and choose *Owl.mp*. You will be asked if you want to save the file from the previous exercise, but you need not do so.
2. Under the *Model* menu select *Populations*. Select each population in turn and record the *Carrying capacity* for each population in Table 12.3.
3. Calculate the amount of money required for restoring each population; do so by multiplying the *Carrying capacity* of each population by $1000. Fill in the amounts in the appropriate column of Table 12.3. The first population is filled in as an example.
4. Choose, at random, a subset of populations that you will restore with your $500,000. In the column marked Scenario A in Table 12.4, put a checkmark next to the subpopulations you have chosen to restore. Be certain that the total restoration cost is $500,000 or less.
5. For each of the populations you have selected, in the *Populations* window, change the *Growth rate* from 0.827 to 1.01. Click OK to exit this window.
6. Now you are going to run the simulation; under the *Simulation* menu select *Run*, or press *Ctrl-R* on your keyboard. The model will run much faster if you view the text rather than the map. To do so, click on the leftmost icon in the simulation window as the simulation is running. At the bottom right corner of this window you will see a message when the simulation is complete. At this point you may close the window by clicking on the X in the upper right corner of the window.
7. Under the *Results* menu, select *Trajectory Summary*. Here you can view the population growth curve of the metapopulation as a whole (referred to as *Population 0*). The time scale is in 1 year time steps. Print the plot of the *Trajectory Summary* by clicking on the printer icon.
8. You can view the numbers corresponding to the plot by clicking the *Show Numbers* button in the upper left corner of the *Trajectory Summary* window. You can return to the plot by clicking the *Show Plot* button, also in the upper left corner of the *Trajectory Summary* window.

Table 12.3. Carrying capacities and cost of restoring 22 populations of the spotted owl in California.

Population	Carrying Capacity	Restoration Cost ($)
N. Monterey	100	100,000
S. Monterey		
C. Alto		
S. Lucia		
S. Madre		
S. Raphael		
S. Ynez		
Pinus		
Tecuya		
Co.		
Tehachapi		
Pelona		
S. Gabriel		
S. Bernadino		
S. Jacinto		
Thomas		
S. Ana		
Palomar		
Black		
Volcan		
Cuyamaca		
Laguna		

9. Under scenario A in Table 12.5, record the final *Average Abundance* of the metapopulation (*Population* 0) and the restoration cost. Exit the *Trajectory Summary* by clicking on the X in the upper right corner of the window.
10. You will now try four other strategies for restoration. Select the populations you will restore based on the criteria specified below for each scenario.

   Scenario B: only the largest populations
   Scenario C: only the smallest populations
   Scenario D: a geographically spread subset of populations
   Scenario E: a group of populations that are all in 1 area.

11. Mark in Table 12.4 which populations you select. Make sure you only spend $500,000 or less for each scenario. For each scenario, assign a *Growth rate* of 1.01 to the selected populations and a *Growth rate* of 0.872 to the nonselected populations. Be certain that you have changed all of the growth rates from the previous scenario to 0.872. Click OK to exit this window.
12. Run the simulation and print your results by repeating steps 6 through 8. As in step 9, record the average abundance of the population at time steps 0, 5, 10, 15, and 20 in the appropriate column of Table 12.5.

Table 12.4. Restoration scenarios for the spotted owl in California.

	Scenario				
Population	A	B	C	D	E
N. Monterey					
S. Monterey					
C. Alto					
S. Lucia					
S. Madre					
S. Raphael					
S. Ynez					
Pinus					
Tecuya					
Co.					
Tehachapi					
Pelona					
S. Gabriel					
S. Bernadino					
S. Jacinto					
Thomas					
S. Ana					
Palomar					
Black					
Volcan					
Cuyamaca					
Laguna					

Table 12.5. The results of the five restoration scenarios and their costs.

Scenario	Restoration Cost ($)	Final Average Metapopulation Abundance
A		
B		
C		
D		
E		

Questions

1. Compare the final abundance you found in the first part of Exercise A to the abundance found with the five restoration scenarios. Did the money spent on restoration make a difference? Is the metapopulation still declining? Explain.

2. Which management action gave you the best results? Why do you think it did?

3. Which management action was the least effective? Why?

4. Which had a greater positive effect on the metapopulation abundance, restoring populations close to each other or those separated from each other? Why would the spatial arrangement of the restored populations make a difference?

5. Based on your conclusions, what suggestions would you give to those responsible on how to carry out conservation efforts with the money available to preserve this species?

> Your lab report should include the following:
>
> 1. An abundance table and answers to questions 1 through 5 for the initial conditions section of Exercise A
> 2. An abundance table and answers to questions 1 and 2 for the effects of dispersal section of Exercise A
> 3. A restoration costs table (Table 12.3), a scenario table (Table 12.4), an abundance table (Table 12.5), 5 population trajectory graphs and answers to questions 1 through 5 for Exercise B

References

LaHaye, W. S., R. J. Gutierrez, and H. R. Akçakaya. 1994. Spotted owl metapopulation dynamics in southern California. *Journal of Animal Ecology*, 63: 775–785.

Thomas, C. D. 1991. Spatial and temporal variability in a butterfly population. *Oecologia*, 87: 577–580.

Laboratory 13
Biodiversity's Biggest Threat: Human Population Growth

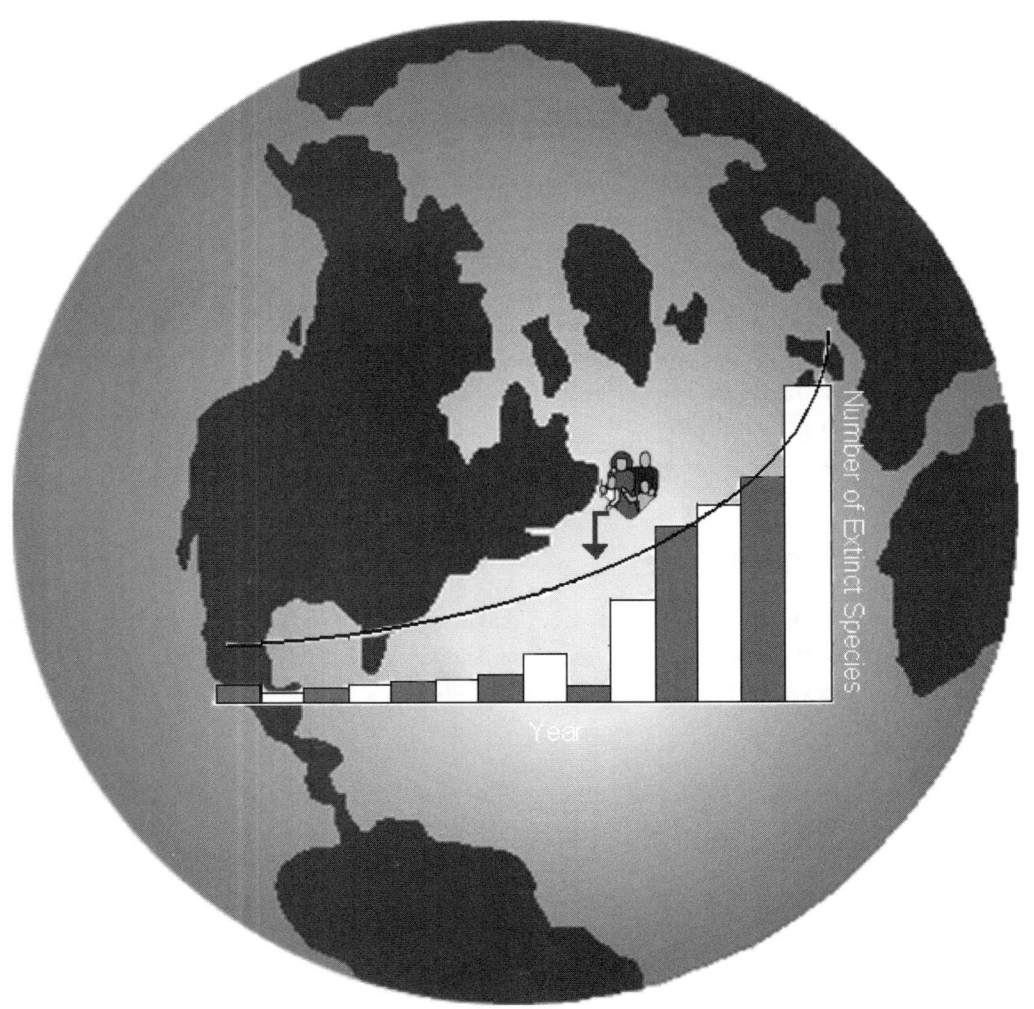

The increasing human population and the rate at which individuals consume resources have important implications for the long-term health of the plants and animals of the planet Earth. In this laboratory you will explore the consequences of human population growth and per capita consumption rate and will examine potential management strategies for controlling the growth rate of humans.

Introduction

If you watched a news program on television last night, you may have seen a story about pollution in a local area. If you read a newspaper this morning, chances are you saw an article on a food shortage in some part of the world. Listening to the radio yesterday, you probably heard about the high levels of crime in this country's urban areas. A magazine that you read recently may have contained an article concerning the plight of endangered species on this planet. All of these issues have one common link, the ultimate cause of each: the growth of the human population.

Never before has a single species had such a huge and devastating impact on the global environment in such a short period of geological time. It is important for those entering the fields of conservation biology or environmental studies to understand the repercussions of human population growth, and the increasing rate of resource consumption, on the Earth's ecosystems and on the standard of living that future generations will experience. There can be no long-term remedies to any environmental or socioeconomic problems if the present rate of growth continues.

The human population is a prime example of a population increasing in an exponential fashion (Figure 13.1). In 1776, when the Declaration of Independence was being signed in Philadelphia, there were 770 million people on the surface of the earth. By 1910, 134 years later, the population had more than doubled so that an additional 1 billion people were alive. By the late 1950s the world's population increased to 2.5 billion people. Since then, the population has more than doubled, with now 5.97 billion, a number difficult for many to fathom.

For most of human history, population growth has been slow and rather steady. However, with the advent of modern medical practices, agriculture, and sanitation methods, the average person lives longer and more people survive to reproduce. These advances have removed many limits to population growth that once kept it relatively stabilized. The medical advances of this century have brought infectious diseases such as diarrhea, pneumonia, tuberculosis, and others under control in the developed parts of the world. By improving health and nutrition, humans now live longer than they once did. In many countries farms are able to produce excess food to support a growing population. Waste management and sewage systems have helped us to grow without poisoning the water supply. The effects of density-dependent factors that might normally have stabilized the population have been reduced.

Although it may not be obvious to people in the developed countries of the world, the present rate of growth not only threatens the ecosystems but also the human standard of living. Humans cannot forever escape density dependence. It is not possible to indefinitely to increase the world's carrying capacity. As human densities continue to increase, starvation, disease, and overcrowding will also increase. At the present time at least 1 billion people are already suffering from malnutrition and starvation. This number can only increase as the population continues to grow and more and more resources are consumed.

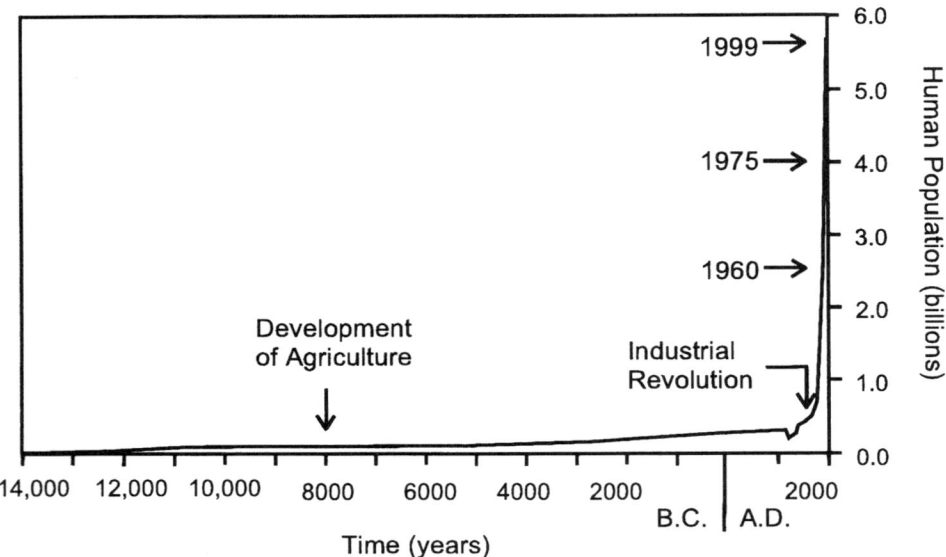

Figure 13.1. The growth of the human population over time. The slight dip between 1347 and 1351 A.D. represents the death of 25 million people as a result of the bubonic plague. (After Starr and Taggart 1995.)

How Many People?

What is the Earth's carrying capacity? How many people can the planet sustainably hold? This has been addressed by many ecologists; yet few agree on a number. In fact, the estimates have ranged between 1 billion and 1 trillion people! This huge difference depends on the standard of living of the average person, the amount and type of resource consumption, advances in agricultural production, and other services. For example, if we were to extend the high-consumption American lifestyle to the nearly 6 billion people on Earth, resources would be quickly depleted, and it would not be long before the high standard of living plummeted. At that consumption level, the planet would have a very small carrying capacity. On the other hand, with the improvement of agriculture and a change in the lifestyles of people in developed nations, the Earth might sustainably hold many more people than are alive today.

A Human-Dominated Planet

If you consider the history of the Earth and of living things, you will note that humans have only been a part of the system for a very short time. Yet humans have dominated the planet and modified the environment like no other species. There is no place on the surface of the globe that has not been affected by human activities in some way. This has resulted from the growth of the human population and also the growth in the amount and variety of resources that humans use.

Through agriculture, industry, hunting, and fishing and through international commerce the human population has attained its present population size and consumption levels. At the same time, these enterprises have resulted in the transformation of natural areas, redistribution of species through biological invasions, and the loss of many species through hunting, fishing, and forest extraction. These things have in turn affected the functioning of ecosystems and the natural cycling of nutrients such as carbon, nitrogen, water and other elements (Figure 13.2). We are facing irreversible losses in biodiversity and are driving changes in the world's climate.

The Earth's ecosystems are definitely changing under human influence, but how will that affect us? At the present time 43% of the Earth's terrestrial surface has lost some capacity for human services, be it agriculture, grazing, or other benefits. The Earth is beginning to lose ecosystem services in many areas. Lands that were once productive are being converted to deserts; wetlands that help purify water are being lost; and pollinators necessary for agricultural production are faltering in many areas. We are losing populations and species that provide us with food or potential new medicines through fishing, hunting, and forest clearing.

Overconsumption

In general, developed countries are growing at a slower pace than poorer, less industrialized nations. Does this mean that the responsibility to the planet's ecosystems belongs to the poorer countries of the world? The lower growth rate in more developed countries does not mean that countries like the United States have less of an impact on their environment. In fact, wealthy developed nations consume more resources and produce more waste per individual than do developing countries. People in the United States consume between 10 and 30 times more resources than people in developing countries. From a global viewpoint, therefore, concern with the consumption and rate of growth of the developed nations is most urgent. Each person added to the United States by birth or immigration will have a greater impact on the environmental problems of the world than a birth in an area such as Bangladesh or Madagascar.

The World Wildlife Fund (Living Planet Report 1998) estimated the average individual consumption pressure (i.e., a measure of the burden placed on the environment of an average individual of the world). The average consumption pressure per individual is illustrated for several countries in Figure 13.3. Taking into account the population size of each country, an estimate of the consumption pressure for each country can also be expressed. As Figure 13.4 illustrates, nations that have a higher standard of living and/or are quite large in population size have the most effect on the environment.

Most developing nations recognize the problems of overpopulation because the people there feel the effects directly. Starvation, malnutrition, and unsafe drinking water are prevalent in these areas. Many of these countries have taken steps to reduce population growth through education and family planning. Resource consumption per individual in these areas and in the developed nations is increasing, however, giving everyone a larger footprint. This increase may be more difficult to control than population growth.

176   Laboratory 13

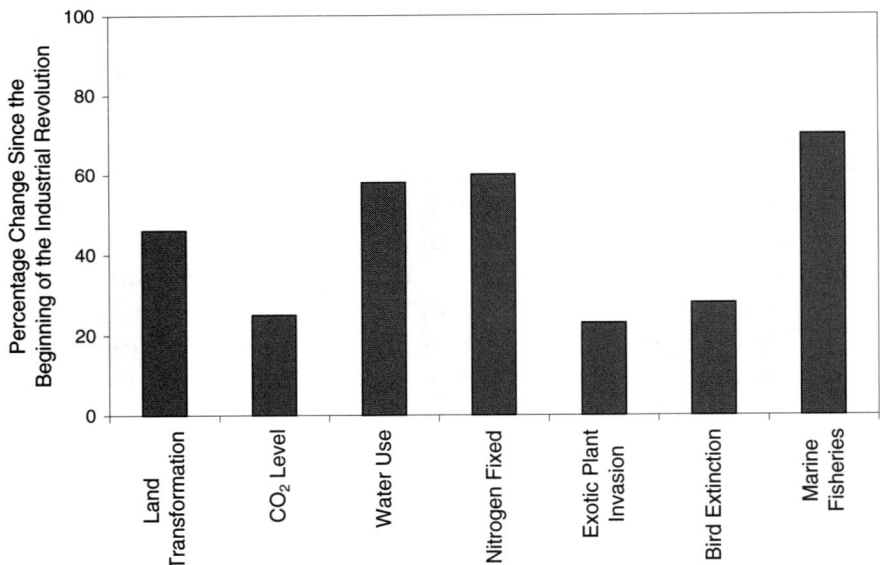

Figure 13.2. The human-induced increases in the amount of land altered, of CO_2 present, of freshwater used, of nitrogen fixed in the soil, of exotic plant invasions, of bird extinctions, and of marine fisheries exploited. (After Vitousek et al. 1997.)

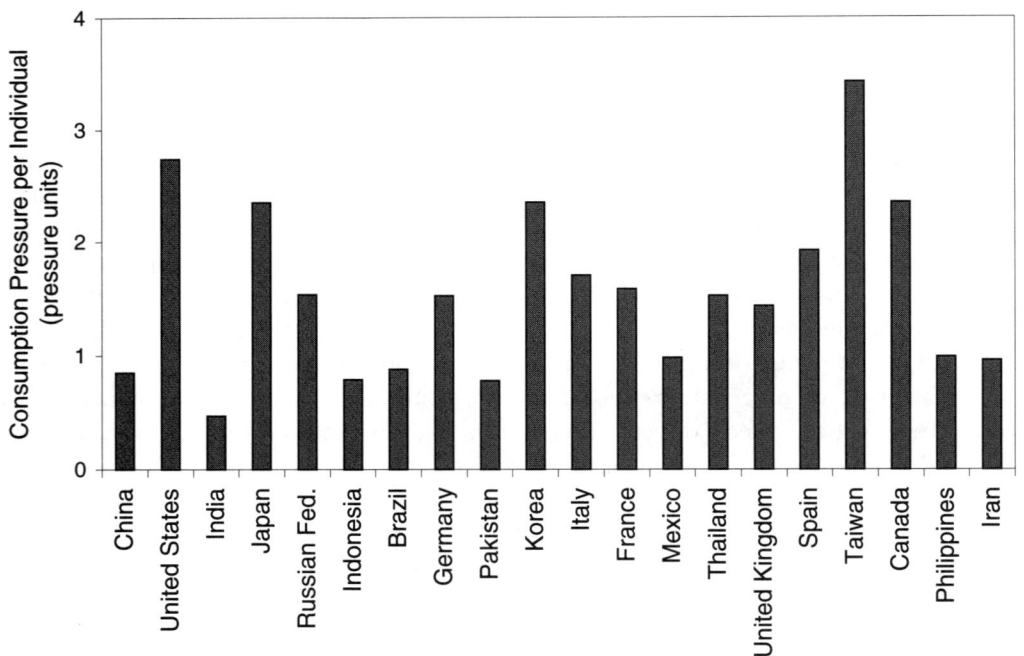

Figure 13.3. The average consumption rate per individual for some of the world's most populous countries. (Adapted from the Living Planet Report 1998 by World Wildlife Fund.)

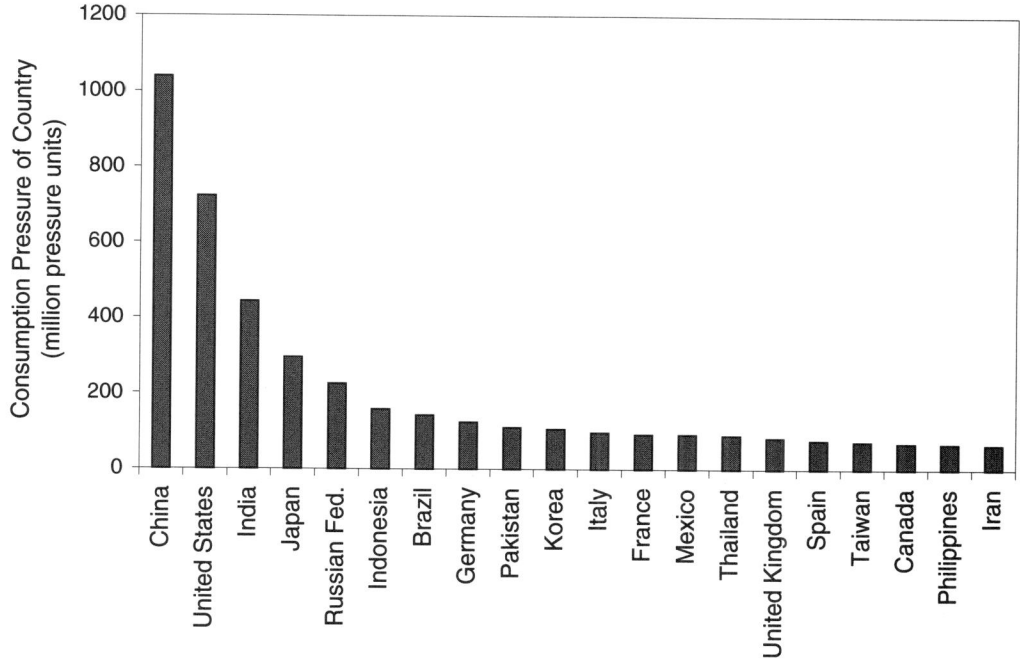

Figure 13.4. A measure of the burden placed on the environment by each country. Consumption pressure is based on an average rate of consumption per person and the total population of the country. (Adapted from the Living Planet Report 1998 by World Wildlife Fund.)

Declining Growth Rates?

In 1988 the growth rate of the world population was 1.7% and the average woman raised 3.6 children. In 1998, the growth rate declined to 1.4% and the average woman was expected to raise 2.9 children. This decline in vital rates is occurring in both the developing and developed nations of the world. The fecundity of the developing world declined from 4.9 to 3.8 (more than one whole child per average woman). Significantly, in the rich, consumptive world of the developed nations, fecundity dropped from 1.9 to 1.6. In Europe the population growth has stopped and has actually started a slow decline. Some people have taken this as dire news, as a sign that the population is in trouble. Instead, this may be very beneficial to the world ecosystem because of vast overpopulation of the continent and because of the large ecological footprints of people in the European nations.

Despite these trends, the world population is still increasing. These changes have only moved the date that the world is expected to reach 8 billion from the year 2019 to 2024. Unless people's standards of living are severely altered, most scientists agree that today's population size is not sustainable forever even if all future growth is halted.

Exercise A: Human population growth

Background

In the exercises in this lab, you will be using an age-structured model with females only to model the human population of the United States. In many studies, demographers (those who study populations) only consider the female component of a population. This is because males rarely limit the reproductive rate of a population. If you were to increase the number of males in the human population, this action would not increase the number of babies born. On the other hand, if you were to add more females we would expect an increase in the number of babies born in each time period. To approximate the *total* United States population you should double the abundance in the model. Interestingly, at birth, there are more males than females in a ratio of 1.04:1.00 (males to females). This ratio stabilizes at 1.00:1.00 at about age 25, and then by the age of 65 there are fewer males than females in a ratio of 1.00:1.20.

Women in the United States reach reproductive age between ages 10 and 20. A small fraction of women under 20 actually give birth; thus the model has low fecundity value in age class 10 to 19. By the age of 50 most women have reached menopause and are no longer able to bear children. Remember that the values in the model represent the number of daughters born to each female, per 10 year period. For each age class there is a chance that an individual will not survive past the next 10 years (age classes are 10 year intervals). The survival rate generally decreases as humans get older. After the age of 40, cancer, heart disease, and other causes of death begin to take their toll on the population.

Exercises

1. Open the program RAMAS EcoLab (double-click on its icon with your mouse).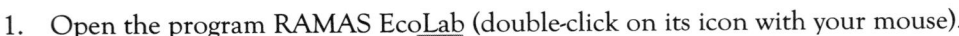
2. Click on *Age and Stage Structure*.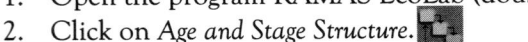
3. Under the *File* menu, select *Open* and choose *USPop.st*. This file contains an age structured matrix of the United States female population in 1993 based on data from the U.S. Census Bureau.
4. Under the *Model* menu select *General Information*. Notice that the *Duration* is set at 5. Which means that the program is set to simulate 5 time steps of reproduction. In this exercise, each time step is 10 years long.
5. Also in the *General Information* window, notice that the *Number of Replications* is set at 1. You will run one simulation to view how the size of the United States population is expected to change over time. One replication is sufficient because you will not be using demographic stochasticity. Click OK to exit this window.
6. Under the *Model* menu select *Initial Abundances*. Notice the structure of the age distribution. The two youngest age classes (0 to 9 and 10 to 19) have fewer individuals than the next two age classes (20 to 29 and 30 to 39).
7. Enter the *Initial Abundances* for each age class in the appropriate column of Table 13.1. Click *Cancel* to exit this window.
8. Under the *Model* menu select *Stage Matrix*. Inspect the matrix, and note that the fecundity for women that are 50 or older is 0. By this age, most American women have reached menopause. The fecundity for women ages 10 to 19 is an average of all women and girls in that age class.
9. Enter the fecundities and survival rates for each age class in the appropriate columns of Table 13.1. Click *Cancel* to exit this window.

10. Under the *Model* menu select *Management & Migration*. Observe the number of female immigrants in each age class every 10 years. Click on each successive *Introduction/Migration* phrase in the box *Management Action* to see how the number of individuals changes for each 10-year age class. The age class and its corresponding number of immigrants are found in the lower right corner of the window.
11. Enter the number of immigrants in each age class in the appropriate column of Table 13.1. Click *Cancel* to exit this window.
12. Now you are going to run the simulation; under the *Simulation* menu select *Run*, or press *Ctrl-R* on your keyboard. At the bottom right corner of this window you will see a message when the simulation is complete. At this point you may close the window by clicking on the X in the upper right corner of the window.
13. Under the *Results* menu, select *Trajectory Summary*. Here you can view the population growth curve. The time scale, like the age classes, is in 10-year time steps. Print the plot of the *Trajectory Summary* by clicking on the printer icon.
14. You can view the numbers corresponding to the plot by clicking the *Show Numbers* button in the upper left corner of the *Trajectory Summary* window. You can return to the plot by clicking the *Show Plot* button, also in the upper left corner of the *Trajectory Summary* window. Exit the *Trajectory Summary* by clicking on the X in the upper right corner of the window.

Table 13.1. Vital statistics of the seven age classes of United States females.

Age Class	Survival Rate	Fecundity	Initial Abundance	Number of Immigrants
0-9	0.9979	0	18,659,000	685,360

Questions

1. According to the *Initial Abundances*, what age range had the most individuals in 1993?

2. According to the *Management and Migration* window, which age range has the highest net immigration every 10 years? Why might this class have the most immigration?

3. Based on the initial abundances and fecundity for age class 20 to 29, how many female offspring will be produced by this group of women in 10 years (one time step)?

4. Notice the number of women and their fecundity for age class 30 to 39. How many female offspring will this group of women produce in 10 years (one time step)?

5. There are fewer women in age class 20 to 29 than in age class 30 to 39, yet the former produces more offspring. Why?

6. If the total population of women in the United States in 1993 was 131,991,000, how many years from 1993 will it take for the United States population to double? (Hint: Under the *Model* menu, select *General Information*. Change the *Duration* to 15, and view the numbers corresponding to the *Trajectory Summary*). Remember to report your result in years, because each time step represents 10 years.

7. Based on this model, how many people of both sexes will there be in the United States by the year 2013?

8. The United States Census Bureau predicts that the United States population will be at 305,112,000 by the year 2013. How does this value compare with your answer to question 7? Be specific.

9. The data in the file *USPOP* was based on the Census Bureau information for 1993. Why does their population estimate for 2013 disagree with yours? (Hint: Do you think the Census Bureau applies the exact same values for birth rate, migration, and survival to each time step?)

Exercise B: The effectiveness of immigration control programs

Background

Recently in the United States Congress there has been concern about population growth in this country and the impact that overcrowding will have on the nation's economy in the next century. There is speculation that health care, unemployment, and social services will all be affected and that, in short, the quality of life in America will decline. One proposal for slowing the rapid increase of the population is to enact stricter immigration laws.

In this section, you will decrease the net immigration rate by 50%. Such a huge cut to current immigration rates would not likely pass as legislation in the United States Congress. The percentage would probably be much lower if such a law were to be enacted.

Exercises

1. Under the *Model* menu, select *Management & Migration*.
2. Click on each successive *Introduction/Migration* phrase in the box *Management Action* to see how the number of individuals changes for each 10 year age class. The age class and its corresponding number of immigrants are found in the lower right corner of the window.
3. Divide each of the migration values by 2, and enter the new value in its place. Do this for all age classes. By doing this you have effectively reduced immigration by 50%. Click OK to exit this window.
4. Now you are going to run the simulation; under the *Simulation* menu select *Run*, or press *Ctrl-R* on your keyboard. At the bottom right corner of this window you will see a message when the simulation is complete. At this point you may close the window by clicking on the X in the upper right corner of the window.

5. Under the *Results* menu, select *Trajectory Summary*. Here you can view the population growth curve. The time scale, like the age classes, is in 10-year time steps. Print the plot of the *Trajectory Summary* by clicking on the printer icon.
6. You can view the numbers corresponding to the plot by clicking the *Show Numbers* button in the upper left corner of the *Trajectory Summary* window. You can return to the plot by clicking the *Show Plot* button, also in the upper left corner of the *Trajectory Summary* window. Exit the *Trajectory Summary* by clicking on the X in the upper right corner of the window.

Questions

1. How many fewer women will there be, compared to the original scenario in Exercise A, after 50 years of this immigration program?

2. This program would reduce the population by what percentage compared to the original scenario?

3. How does the pattern of growth compare to the original scenario in Exercise A (rate of change and shape of curve)?

4. Do you think that a 10% reduction in migration would have a significant effect on the population size in the next 50 years? Why or why not?

Exercise C: Fertility control programs

Background

In 1990 the population of China was estimated to be about 1.13 billion, roughly one-fifth of the world's population at that time. The number of new babies born there each year, about 22 million, is equal to the populations of New York, Chicago, and Los Angeles combined. For many centuries, China went through periods of rapid growth and then sudden decline due to natural disasters, famine, and political upheavals. However, from 1950 to 1980 the death rate dropped from 20 per thousand to 8 per thousand and the average life expectancy increased from forty-seven to seventy years, producing an extended period of rapid growth. This trend attracted the attention of demographers and politicians all over the world. They expressed concern that China would not be able to feed or shelter its people in the next century.

In the 1980s the Chinese government implemented a "One Child Policy" to halt its rapidly expanding population. The policy was designed to hold the population to 1.2 billion by the year 2000. Under the guidelines of this program couples were rewarded for limiting themselves to one child and penalized, with fines or other economic sanctions, for second or additional births. The implementation of this program gained much attention internationally as other countries observed the social, economic, and demographic effects.

However hypothetical this may sound, consider a similar program for the population of the United States, or alternatively, imagine that social and economic factors cause a decline in the birth rate. American couples on average have 2.1 children during their lifetimes. Figure 13.5 illustrates the total number of births annually for the last seventy years. In the next exercise you will simulate the effects of reducing fecundity, the birth rate, by half.

Exercises

1. Under the *File* menu, select *Open* and choose *USPop.st*. You will be asked if you want to save the file from the previous exercise, but you need not do so.
2. Under the *Model* menu select *Stage Matrix*. Divide each of the fecundity values by 2, and enter the new values into the matrix. Be careful that you do not divide the survival values by 2. Click OK to exit this window.
3. Now you are going to run the simulation; under the *Simulation* menu select *Run*, or press *Ctrl-R* on your keyboard. At the bottom right corner of this window you will see a message when the simulation is complete. At this point you may close the window by clicking on the X in the upper right corner of the window.
4. Under the *Results* menu, select *Trajectory Summary*. Here you can view the population growth curve. The time scale, like the age classes, is in 10-year time steps. Print the plot of the *Trajectory Summary* by clicking on the printer icon.
5. You can view the numbers corresponding to the plot by clicking the *Show Numbers* button in the upper left corner of the *Trajectory Summary* window. You can return to the plot by clicking the *Show Plot* button also in the upper left corner of the *Trajectory Summary* window. Exit the *Trajectory Summary* by clicking on the X in the upper right corner of the window.

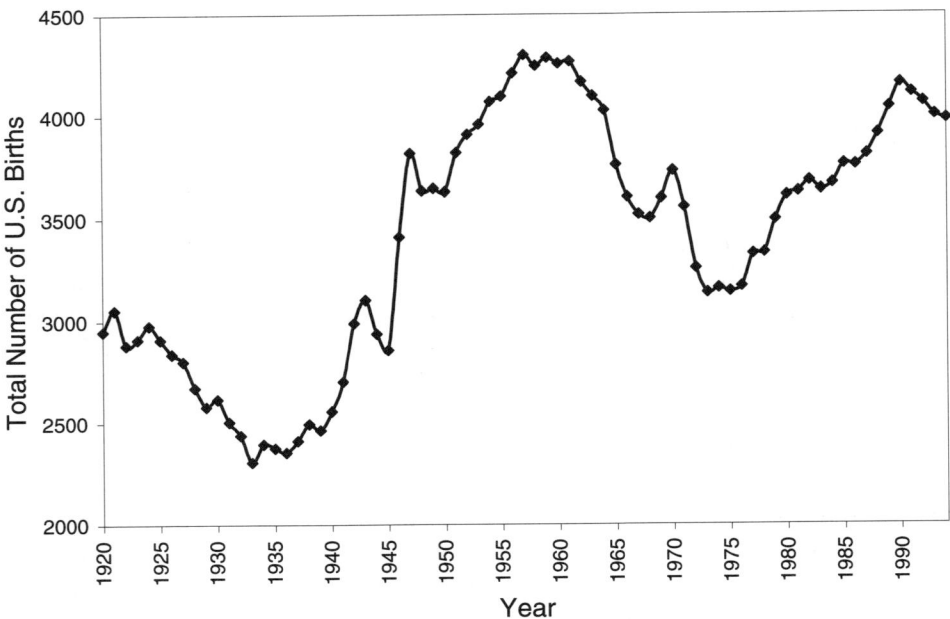

Figure 13.5. A graph of the total number of births per year in the United States. The large peak between 1947 and 1965 is the so-called baby boom.

Questions

1. How many fewer women will there be, compared to the original scenario in Exercise A, after fifty years of this fertility rate?

2. A 50% reduction in fertility would reduce the population by what percentage compared to the original scenario?

3. How does the effectiveness of this fertility program compare to that of the immigration program, in limiting the United States population's growth rate?

4. Even though this program enforces approximately *one* child per *two* individuals, why is there still an increase in the population for the first 20 years?

5. What would be some of the arguments *for* and *against* effecting such a reduction in fertility as a governmental policy?

6. Consider what would happen to the growth rate of this country's population if the following scenario were to take place. Rather than lowering the average birth rate by 50%, women instead delayed reproduction until at least the age of 30 (all other factors being equal).
   a. Which of these two conditions is more effective at lowering the population's growth rate?

   b. Which of these conditions do you feel would be more acceptable as a governmental policy? Why?

Your lab report should include the following:

1. One vital statistics table (Table 13.1), 1 population trajectory graph, and answers to questions 1 through 9 for Exercise A
2. One population trajectory graph and answers to questions 1 through 4 for Exercise B
3. One population trajectory graph and answers to questions 1 through 6 for Exercise C

References

Starr, C. and R. Taggart. 1995. *Biology: The Unity and Diversity of Life.* Wadsworth Publishing Co., Belmont, CA.

Vitousek, P. M., H. A. Mooney, J. Lubchenco, and J. M. Melillo. 1997. Human domination of Earth's ecosystems. *Science,* 227: 494–499.

World Wildlife Fund. 1998. *Living Planet Report 1998.* (available on-line at http://panda.org/livingplanet/lpr/index.htm).

Also available on-line: U.S. Census Bureau at http://www.census.gov.

Laboratory 14
The Case of Patrick's Marsh Wren: Making Decisions to Protect Species

The application of conservation biology to find solutions for actual situations is almost always much more complicated than simply applying conservation theory. Conflicting interests, incomplete information, and unclear predictions together produce a situation that can be extremely difficult to resolve.

Introduction

Now that you have been introduced to the theory used in conservation and population biology, it is your turn to make the difficult decisions often necessary in conservation biology. Complications arise when opposing interest groups such as developers, environmental activists, and state and federal agencies must agree about which, if any, actions should be taken in a particular situation. Each group presents data and arguments that support its position, often resulting in several biased and incomplete depictions of the real case. Making well-informed decisions and recommendations can be difficult. Environmental activists promote the well-being of a species, regardless of the inconveniences it necessitates; developers stress the importance of human interests over the survival of a species. Deciding which actions are the best compromise for all positions can be extremely difficult. In this laboratory, we are going to use many of the concepts presented in previous labs to make some recommendations in a hypothetical case in the Central Valley of California.

Your job is to apply the analytical techniques from the previous chapters, especially those from metapopulation dynamics, to the given scenario. Depending on which side of the case you are on, you will try to find data that support your position and data that weaken that of your opponents. A resolution can only happen when all sides come to an agreement. Remember that incomplete information, conflicting observations, and different assumptions complicate many legal battles involving endangered species. The more serious you are about exploring the dynamics at play in this scenario, the more fully you will understand why there are conflicts in the real world.

Conservation Planning

One question we have not addressed in the previous laboratories is why we conserve in the first place. To many students in life sciences and environmental studies, the need to preserve the environment and species is undeniable. However, there are still many who need to be convinced that they should be inconvenienced for the future of a frog, a beetle, or a snail. As conservation biologists, we need to be able to make well-informed decisions for the conservation movement.

Several arguments have been introduced in support of biological conservation. The first argument is strictly economic. Humans have and will directly benefit from natural species and their components. As omnivores we directly consume many types of plants and animals. Timber provides building material, paper, and fuel. Native species also have provided a vast arsenal for our war on disease. Commonly used medicines such as salicylic acid (aspirin) were first isolated from species growing in their native habitat. Species such as the Madagascar periwinkle have helped conquer deadly diseases. Many argue that the world's species hold the key to defeating even our most devastating diseases.

A second argument is that of indirect benefits or services provided by the environment. Each species plays an integral role in its community and has potential economic value. An intact community is necessary for an ecosystem to function properly to help filter water and cleanse the air. There are also other indirect benefits such as the booming industry of ecotourism. Ecotourism depends on preserving intact habitats and natural areas for people to visit. Economic arguments are often used; products provided by the environment, however, generally do not immediately offset the economic costs of preserving habitat. In addition, not every species will produce a marketable commodity and only very few will be found to have valuable pharmaceutical qualities.

A third argument is that it is our moral responsibility to respect and preserve nature. Every species is unique, by definition. This argument is ethical as opposed to economic. Some argue that all species have an intrinsic right to life and humans benefit in a spiritual way from the intact natural world (the wonder we feel when looking at mountains is often given as an example). Humans do not have the right to destroy the environment on an unprecedented scale.

These arguments vary in their effectiveness depending on the species examined and the audience. It is always important to remember that not everyone believes it is critical to preserve biodiversity. These underlying beliefs shape the actions and arguments that arise when development, or "progress", collides with preservation and conservation. There are consequences and costs to all the choices in any conflict. In the following exercise, you will be given a chance to debate some of these issues and investigate their consequences using a hypothetical conservation case, the Patrick's marsh wren.

Patrick's Marsh Wren

Patrick's marsh wren (*Cistothorus palustris dawsonii*) is found only in a few small populations in the southeastern Central Valley in California. It is an isolated population of the common marsh wren (*Cistothorus palustris*). However, because of differences in appearance and behavior, many ornithologists believe Patrick's marsh wren should be a distinct subspecies. Patrick's marsh wren can be easily recognized by the bar pattern on its belly and its distinct chirping call. The local chapters of several environmental organizations including Peterson's Birders, the Sequoia Club, and Defenders of Nature are currently lobbying Congress and California Fish and Game to recognize Patrick's marsh wren as a separate subspecies and protect it under the Endangered Species Act. Very little genetic information has been obtained from individuals of the Patrick's marsh wren population. We can only use morphological traits (or appearance) and behavioral traits to support Patrick's marsh wren status as a separate subspecies.

Patrick's marsh wren lives in a very limited geographic area and within a narrow habitat range. It nests in wetland reeds and grasses. It feeds on insects that accumulate around standing water and depends on the water to protect against nest predation by coyotes and raccoons. Patrick's marsh wren migrates to Baja California in the winter but returns to the Central Valley each spring to breed and raise young. This wren has very high site fidelity; individuals tend to return to the same nesting sites year after year. When individuals mature, they are not known to disperse more than 35 km (21.8 miles) to new breeding grounds.

The Scenario

Green Valley Ranch, Inc., a major beef producer, owns much of the land surrounding the town of Yokeston, near the foothills of the southern Sierra Nevada Range. The company has recently increased the size of its herds and now needs more pasture to allow them to graze. The ranch's property runs along the Rio Niebla, a major watershed of the Sierra Nevadas. There are several dams upstream from Yokeston, and flood-control dikes and levies contain much of the river in the Central Valley. One of the few lengths of the river that is still untouched is on the Green Valley Ranch and surrounding cattle ranches. Because of the intense seasonal flooding and semipermanent wetland, the land along this stretch of river is not suitable for grazing. To begin using this land, Green Valley Ranch wants to erect a dike along the river to prevent seasonal flooding. The area they wish to reclaim is one of the few remaining habitats of the Patrick's marsh wren.

Bill Kidd, the owner of Green Valley Ranch, petitioned the state for permits to build the dike on his property. Landowners have only recently needed special permits for erecting flood-control structures on their property. Mr. Kidd argues that because his family has owned the land for several generations, they should fall under a grandfather clause. A grandfather clause allows individuals to abide by previous regulations if they owned a property before a new law is instituted. Green Valley is fairly environmentally sensitive, and it does not remove water from the river for irrigation or use pesticides on any part of its property. Mr. Kidd argues the damage done to the watershed by creating a small dike is minimal.

The state asked the California Fish and Game Department to conduct an environmental impact study of the area. When the local environmental groups heard about the proposal, they joined forces to prevent Green Valley Farm from redirecting any water. Because of the pending legislation about the species status of Patrick's marsh wren, they argue we should invoke the "takings" law from the Endangered Species Act. The takings law prevents landowners from disturbing endangered species by capturing them, negatively altering their habitat, or interfering with them in any harmful way. Because the populations of Patrick's marsh wren are small and isolated, removing any habitat from the species could potentially cause all of the populations to collapse. Environmental groups argue the land in question, Philip Flat, is integral to the stability of the 3 other populations in the area. They believe Philip Flat provides a dispersal path between Three Mile Marsh, the large population to the northwest, and two others, Delinger Meadow to the south and Castleton Creek to the northeast.

After reading the complaints from the environmental lobby, Fish and Game stalled the permit application to conduct an in-depth environmental impact report. It wanted to perform a population viability analysis for Patrick's marsh wren. After waiting an additional 6 months, Mr. Kidd filed a lawsuit against the state for stalling and lack of precedent to prohibit development. His increased herd was overgrazing the available land and neither he, nor his cows, had the patience to wait for the state. He had heard that deliberations on endangered species status and protection could last well over a decade. Because the populations in question have not yet been recognized as a distinct subspecies, Mr. Kidd feels the state interference is without basis.

The state was unable and unwilling to put up funds to purchase the contested land from Mr. Kidd. In the ensuing period of indecision, Mr. Kidd attempted to broker a land-swap settlement. In return for permission to build the proposed dike, he would remove an old and disintegrating dike in the southern part of his property (Figure 14.1). By removing the dike, he would increase the wetland area available for the small population of the Patrick's marsh wren in Delinger Meadow. The result would be a redistribution of the available habitat but not a reduction in total area. A highway separates this land from the main pastures of Green Valley Ranch, so the ranch sees this proposal as a win-win situation. Green Valley Ranch is also willing to agree to a conservation easement, an agreement that would prohibit future development on the land.

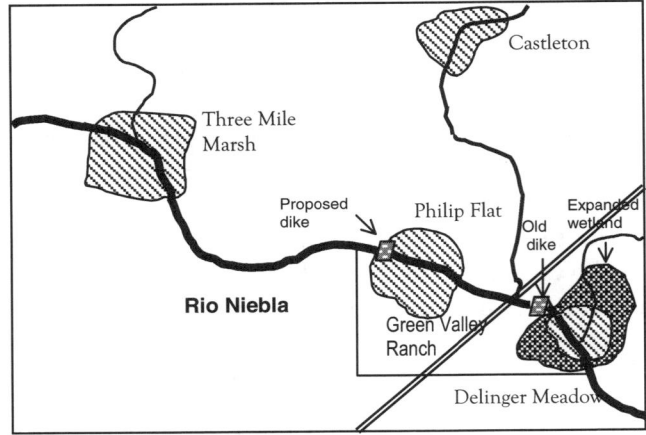

Figure 14.1. Map of the Green Valley Ranch and the surrounding area.

Exercise: Evaluating options for Patrick's marsh wren

Use the *Multiple Populations* program in RAMAS EcoLab to open the *Wrenpop1.mp*, *Wrenpop2.mp*, and *Wrenpop3.mp* files with the population data for Patrick's marsh wren. When you run the simulation, click on the *text* icon in the upper left of the screen where the simulation is running. This will allow the simulation to run much faster. When you make changes to the population parameters in each of the files, use the *Save as* function in the *File* menu to save the changes under a new name to make sure you keep the values in the original files.

The first file, *Wrenpop1.mp*, contains the information for scenario A, which is the current state of 4 small populations. There are estimates for the population size, growth rate and location of each of the known populations. The second file, *Wrenpop2.mp*, depicts the land-swap plan proposed by Mr. Kidd. In scenario B the Philip Flat habitat has been removed and the carrying capacity of Delinger Meadow was increased by the same amount as the pervious carrying capacity of Philip Flat. Assume that only a portion of the animals from Philip Flat population are able to successfully disperse to Delinger Meadow. You will notice that the initial population size has been increased by 20 individuals. In the third file, *Wrenpop3.mp*, the Philip Flat population has been removed but added no area (or carrying capacity) was added to any of the remaining populations.

The class will be divided into 4 teams representing different aspects of the case: (1) the lobby for Green Valley Ranch, (2) the lobby for the environmental groups, (3) representatives of the California Fish and Game Department, (4) the state supreme court judges. Each team is to evaluate the importance of the Philip Flat population in the overall stability of the species. Guidelines are provided for the strategy of each team. Determine whether the settlement offered by Mr. Kidd is a reasonable alternative. Does it provide a more stable alternative than the single small Philip Flat population? What role does the uncertainty about different population parameters make?

After each team makes its evaluation, you will present your results to the class. Try to make as strong a case as possible for your particular position. The state wants to present the case as objectively as possible. Green Valley Ranch wants a quick settlement. The environmental groups want neither a permit issued nor any alteration of the habitat. Once these views are presented, the group of judges will have to decide what is the best course of action. The team of judges will decide on the structure of the hearing/trial and inform all of the teams as soon as possible so that they may prepare appropriately.

Two useful sources of information, on the World Wide Web, about the Endangered Species Act (ESA) and habitat conservation are:
Pro ESA:
www.audubon.org/campaign/index.html
Opposed to the ESA:
www.nesarc.org

FACT SHEET FOR PATRICK'S MARSH WREN

Range Patrick's marsh wren is found only in the southern part of California's Central Valley. However, it is still officially regarded as a subpopulation of the marsh wren, a species found over much of North America. All populations of the marsh wren winter in northern and central Mexico.

Diet Patrick's marsh wren feeds mainly on insects found within and on the edges of wetland areas. Spiders make up a substantial part of their diet.

Population estimates The 4 known populations of Patrick's marsh wren have been sampled.

	Abundance
Three Mile Marsh	75
Philip Flat	30
Castleton Creek	23
Delinger Meadow	30

Note: Populations are assumed to be near half of the carrying capacity. The population sizes were used to estimate the carrying capacity for each population. This may be an overestimate or an underestimate of carrying capacity.

Growth rate The lowest growth rate recorded for any population was 1.02 with a standard error of 0.2. Other studies have measured the growth rate at 1.05 and 1.08. The model assumes an intermediate growth rate of 1.05 that is close to an average, but there is little information to support this assumption.

Dispersal ability Although able fliers, Patrick's marsh wren seems to disperse only to very nearby breeding grounds. Most adults breed in or near their birthplace, but a few disperse to nearby populations. The farthest any known individual has dispersed is 35 km (21.8 miles), but this is based on very few observations, so in reality the maximum distance may be greater. The average dispersal distance is an estimate based on information from other similar species.

Team 1: Green Valley Ranch

Objective: Your objective as an advocate for the interests of the Green Valley Ranch is to resolve this issue in as short a time as possible with minimal cost to your client.

Best Case Scenario: The best situation would be for the court to dismiss the concerns of the environmental groups, and for California Fish and Game to immediately issue the permit for dike construction.

Worst Case Scenario: The worst situation would be if the dike construction was prohibited and no new construction of any kind was allowed on the property. Recent disputes indicate this outcome is unlikely, but estimating the future population growth with no settlement may make a strong case for the importance of this population.

Strategy

Argue that there is little harm to the wren metapopulation if the dike is built, supported by population viability analysis, but be willing to compromise by destroying the old dike and including a conservation easement. Stress the economic implications the worst case scenario option would have for the ranch. Use your results (and answers to the questions below) from the models to support your arguments.

Procedure

Each team member should run the models *Wrenpop1.mp*, *Wrenpop2.mp*, and *Wrenpop3.mp*. From your results you will calculate the percentage change in population size for each of the scenarios by dividing the change in abundance by initial abundance. You will also examine the risk of extinction and the risk of declining below 100 or 50 individuals. The extinction risk and risk of decline can be found in the *Extinction/Decline* function under the *Trajectory Summary* menu. Click on the *table* icon on the menu bar. The numbers show the probability of the population falling below each abundance.

Your group will also change the carrying capacity, the growth rates, and the estimated average dispersal distance of the populations to reflect uncertainty in the estimates of the habitat quality. Each time you change these values, you are trying to determine the effect inaccurate information has on your predictions.

Fill in the tables at the end of the laboratory to help summarize the results. Tables 14.8 and 14.9 are optional, but it might strengthen your case to change more than one of the uncertain values in the model (such as carrying capacity and dispersal distance). Keep in mind that your evaluation is supposed to mimic an actual dispute over an environmental issue. In conservation disputes, the answers are rarely clear cut, and often the best solutions represent a compromise that is sensitive to the future of the species in question and the other parties concerned.

Questions

1. Do differences exist between the risks of decline and extinction of the metapopulation under the three scenarios when the original parameters are used? (Hint: compare equivalent risks such as the probability of dropping below 100, 50, and reaching extinction.)

2. Describe the magnitude of difference between the extinction and decline risks of strategy A and B using the original values? Do you think this magnitude is great enough to qualify prohibiting Mr. Kidd from eliminating the Philip Flat population?

3. How much variation is there in the average final abundance estimates? When might the final abundance estimates be useful (think about growth rates)?

4. How do the results change if the growth rate is reduced to 1.02 or increased to 1.08?

5. How do the results differ if the average dispersal distance is reduced to 15 km (9.3 miles) or increased to 20 km (12.4 miles) from the original distance of 18.5 km (11.5 miles)?

6. How does the carrying capacity of Philip Flat affect the future growth rate of the population?

7. What can you infer about the sensitivity of the PVA to errors in the estimates of dispersal and growth rate?

8. What responsibility do you think the state has toward landowners if it is going to limit use of their property? Discuss one reason Mr. Kidd might think it is unfair that he, a single individual, be pressured to take responsibility for the future of a population?

9. Prepare a short statement (2 or 3 paragraphs) describing your argument, its main points, and supporting evidence.

Team 2: Environmental Groups

Objective: Your objective as an advocate for the interests of the environmental groups is to prevent further habitat destruction and possible declines in Patrick's marsh wren.

Best Case Scenario: The best situation would be for the court to prevent the Green Valley Ranch from any alterations to the habitat and be prevented from any other future development.

Worst Case Scenario: The worst situation would be if the dike construction was permitted and no additional considerations for the habitat (such as removal of the old dike and a conservation easement) were included.

Strategy

Argue that there is significant potential harm to the wren metapopulation if the dike is built, supported by population viability analysis. Your position is firmly rooted in the belief that the wren is a unique natural resource that should be protected to the fullest extent allowed under the Endangered Species Act. Use results (and answers to the questions below) from the models to support your arguments.

Procedure

Each team member should run the models *Wrenpop1.mp*, *Wrenpop2.mp*, and *Wrenpop3.mp*. From your results you will calculate the percentage change in population size for each of the scenarios by dividing the change in abundance by initial abundance. You will also examine the risk of extinction and the risk of declining below 100 or 50 individuals. The extinction risk and risk of decline can be found in the *Extinction/Decline* function under the *Trajectory Summary* menu. Click on the *table* icon on the menu bar. The numbers show the probability of the population falling below each abundance.

    Your group will also change the carrying capacity, the growth rates, and the estimated average dispersal distance of the populations to reflect uncertainty in the estimates of the habitat quality. Each time you change these values, you are trying to determine the effect inaccurate information has on your predictions.

    Fill in the tables at the end of the laboratory to help summarize the results. Tables 14.8 and 14.9 are optional but it might strengthen your case to change more than one of the uncertain values in the model (such as carrying capacity and dispersal distance). Keep in mind that your evaluation is supposed to mimic an actual dispute over an environmental issue. In conservation disputes, the answers are rarely clear cut, and often the best solutions represent a compromise that is sensitive to the future of the species in question and the other parties concerned.

Questions

1. Do differences exist between the risks of decline and extinction for the metapopulation under the 3 scenarios, when the original parameters are used? (Hint: compare equivalent risks such as the probability of dropping below 100, 50, and reaching extinction.)

2. How much variation is there in the average final abundance estimates? When might the final abundance estimates be useful (think about growth rates)?

3. How do the results change if the growth rate is reduced to 1.02 or increased to 1.08?

4. How do the results differ if the average dispersal distance is reduced to 15 km (9.3 miles) or increased to 20 km (12.4 miles) from the original distance of 18.5 km (11.5 miles)?

5. How does the carrying capacity of Philip Flat affect the future growth rate of the population?

6. What can you infer about the sensitivity of the PVA to errors in the estimates of dispersal and growth rate?

7. If you are interested in the conservation of a species but uncertain in the estimates of the population parameters, should you support a more conservative or less conservative action plan? What are the risks you face if you base your decisions on overly optimistic estimates?

8. What evidence is there that this metapopulation is a unique subspecies that deserves protection under the Endangered Species Act?

9. Prepare a short statement (2 or 3 paragraphs) describing your argument, its main points, and supporting evidence.

Team 3: California State Fish and Game

Objective: Your objective as a representative of the state of California is to obey the laws at a state and federal level concerning endangered species. This includes preventing harm to any species considered threatened at the state or federal level.

Best Case Scenario: The best situation would be for the court to rule that the Fish and Game Commission acted in accordance with the law by ordering more study before issuing a permit.

Worst Case Scenario: The worst situation would be the Fish and Game Commission is found guilty of obstructing the Green Valley Ranch, and is forced to pay reparations to them for any loss of business revenue and issue an immediate permit.

Strategy

Argue that there is evidence of potential harm to the wren metapopulation if the dike is built, supported by population viability analysis, and that additional study of the situation is warranted before a permit could be issued. Your decision to accept the compromise proposed by Mr. Kidd should be made only after carefully reviewing the results of the models. Use results from the models and answers to the questions to support your arguments.

Procedure

Each team member should run the models *Wrenpop1.mp*, *Wrenpop2.mp* and *Wrenpop3.mp*. From your results you will calculate the percentage change in population size for each of the scenarios by dividing the change in abundance by initial abundance. You will also examine the risk of extinction and the risk of declining below 100 or 50 individuals. The extinction risk and risk of decline can be found in the *Extinction/Decline* function under the *Trajectory Summary* menu. Click on the *table* icon on the menu bar. The numbers show the probability of the population falling below each abundance.

Your group will also change the carrying capacity, the growth rates, and the estimated average dispersal distance of the populations to reflect uncertainty in the estimatesof the habitat quality. Each time you change these values, you are trying to determine the effect inaccurate information has on your predictions.

Fill in the tables at the end of the laboratory to help summarize the results. Tables 14.8 and 14.9 are optional, but it might strengthen your case to change more than one of the uncertain values in the model (such as carrying capacity and dispersal distance). Keep in mind that your evaluation is supposed to mimic an actual dispute over an environmental issue. In conservation disputes, the answers are rarely clear cut, and often the best solutions represent a compromise that is sensitive to the future of the species in question and the other parties concerned.

Questions

1. Do differences exist between the risks of decline and extinction for the metapopulation under the three scenarios, when the original parameters are used? (Hint: compare equivalent risks such as the probability of dropping below 100, 50, and reaching extinction.)

2. How much variation is there in the average final abundance estimates? When might the final abundance estimates be useful (think about growth rates)?

3. How do the results change if the growth rate is reduced to 1.02 or increased to 1.08?

4. How do the results differ if the average dispersal distance is reduced to 15 km (9.3 miles) or increased to 20 km (12.4 miles) from the original distance of 18.5 km (11.5 miles)?

5. How does the carrying capacity of Philip Flat affect the future growth rate of the population?

6. What can you infer about the sensitivity of the PVA to errors in the estimates of dispersal and growth rate?

7. Do you think there is sufficient data to make a decision on either listing this species (providing it with protection) or allowing the development? Explain.

8. What are the implications of the status of the Patrick's marsh wren being considered a separate subspecies or just another population of the marsh wren?

9. Prepare a short statement (2 or 3 paragraphs) describing your argument, its main points, and supporting evidence. Remember Green Valley Ranch brought the lawsuit against you for failing to make a decision in a timely manner on the issue of a dike construction permit.

Team 4: State Supreme Court

Objective: Your objective as the judges hearing this case is to evaluate all of the evidence and decide upon a course of action to resolve the current impasse.

Best Case Scenario: The best situation would be to reach a compromise that allows some of Green Valley Ranch's demands, yet satisfies the Fish and Wildlife Commission's concerns about the protection of a possibly threatened species.

Worst Case Scenario: The worst situation would be if no compromise is agreeable to both sides then the court has to impose the decision upon both parties and probably suffer endless subsequent appeals.

Strategy

Your team has three main responsibilities. First, while the other groups are composing their arguments, decide how the trial/hearing will be structured and inform all teams as soon as possible. Second, evaluate the evidence independently by performing your own population viability analysis so that you are equipped with the knowledge to properly weigh both sides' arguments. Third, you must conduct the hearing (act as a referee if things get heated) and render a verdict. Use results (and answers to the questions below) from the models to support your arguments.

Procedure

Each team member should run the models *Wrenpop1.mp*, *Wrenpop2.mp*, and *Wrenpop3.mp*. From your results you will calculate the percentage change in population size for each of the scenarios by dividing the change in abundance by initial abundance. You will also examine the risk of extinction and the risk of declining below 100 or 50 individuals. The extinction risk and risk of decline can be found in the *Extinction/Decline* function under the *Trajectory Summary* menu. Click on the *table* icon on the menu bar. The numbers show the probability of the population falling below each abundance.

    Your group will also change the carrying capacity, the growth rates, and the estimated average dispersal distance of the populations to reflect uncertainty in the estimatesof the habitat quality. Each time you change these values, you are trying to determine the effect inaccurate information has on your predictions.

    Fill in the tables at the end of the laboratory to help summarize the results. Tables 14.8 and 14.9 are optional, but it might strengthen your case to change more than one of the uncertain values in the model (such as carrying capacity and dispersal distance). Keep in mind that your evaluation is supposed to mimic an actual dispute over an environmental issue. In conservation disputes, the answers are rarely clear cut, and often the best solutions represent a compromise that is sensitive to the future of the species in question and the other parties concerned.

Questions

1. Do differences exist between the risks of decline and extinction for the metapopulation under the three scenarios, when the original parameters are used? (Hint: compare equivalent risks such as the probability of dropping below 100, 50, and reaching extinction.)

2. Describe the magnitude of difference between the extinction and decline risks of strategy A and B using the original values? Do you think this magnitude is great enough to qualify prohibiting Mr. Kidd from eliminating the Philip Flat population?

3. How much variation is there in the average final abundance estimates? When might the final abundance estimates be useful (think about growth rates)?

4. How do the results change if the growth rate is reduced to 1.02 or increased to 1.08?

5. How do the results differ if the average dispersal distance is reduced to 15 km (9.3 miles) or increased to 20 km (12.4 miles) from the original distance of 18.5 km (11.5 miles)?

6. How does the carrying capacity of Philip Flat affect the future growth rate of the population?

7. What can you infer about the sensitivity of the PVA to errors in the estimates of dispersal and growth rate?

8. Do you think there is sufficient data to make a decision on either listing this species (providing it with protection) or allowing the development? Explain.

9. Prepare a short outline describing how you will conduct the hearing/trial including who is allowed to speak and for how long, what evidence will be allowed, and possible decisions. Remember that the Green Valley Ranch brought the lawsuit against the California Fish and Game Commission.

Worksheets

Remember to save your results from each of the simulations with a different name to preserve the original files.

Table 14.1. Simulation with Initial Population Parameters.

(a)

	Initial Population Abundance	Final Abundance	% Population Change [(Final–Initial) / Initial] × 100
Strategy A			
Strategy B			
Strategy C			

(b)

Probability of Decline	Strategy A	Strategy B	Strategy C
>100			
> 50			
0			

Table 14.2. Suppose your estimate of the population growth rate is too low. Change average growth rate from 1.05 to 1.08 in all populations.

(a)

	Initial Population Abundance	Final Abundance	% Population Change [(Final–Initial) / Initial] × 100
Strategy A			
Strategy B			
Strategy C			

(b)

Probability of Decline	Strategy A	Strategy B	Strategy C
>100			
> 50			
0			

Table 14.3. Suppose your estimate of the average population growth rate is too high. Change average growth rate from 1.05 to 1.02 in all populations.

(a)

	Initial Population Abundance	Final Abundance	% Population Change [(Final–Initial) / Initial] × 100
Strategy A			
Strategy B			
Strategy C			

(b)

Probability of Decline	Strategy A	Strategy B	Strategy C
>100			
> 50			
0			

Table 14.4. Suppose you have underestimated the average dispersal distance. Change the dispersal distance in the original files to 20 km.

(a)

	Initial Population Abundance	Final Abundance	% Population Change [(Final–Initial) / Initial] × 100
Strategy A			
Strategy B			
Strategy C			

(b)

Probability of Decline	Strategy A	Strategy B	Strategy C
>100			
> 50			
0			

Table 14.5. Suppose your estimate of average dispersal distance is too high. Change the average dispersal distance in all scenarios to 15 km.

(a)

	Initial Population Abundance	Final Abundance	% Population Change [(Final–Initial) / Initial] × 100
Strategy A			
Strategy B			
Strategy C			

(b)

Probability of Decline	Strategy A	Strategy B	Strategy C
>100			
> 50			
0			

Table 14.6. Suppose you have incorrectly estimated carrying capacity of the different habitats based on current population abundances. There is a possibility that the habitat in the Philip Flat population can support many more animals. Change the carrying capacity (K) of the Philip Flat population from 90 to 130. Note: Strategies B and C will remain unchanged.

(a)

	Initial Population Abundance	Final Abundance	% Population Change [(Final–Initial) / Initial] × 100
Strategy A			
Strategy B			
Strategy C			

(b)

Probability of Decline	Strategy A	Strategy B	Strategy C
>100			
> 50			
0			

Table 14.7. Suppose you have overestimated the carrying capacity of the Philip Flat population. Explore what happens to your predictions if you lower the carrying capacity from 90 to 50 in the Philip Flat population. Note: Strategies B and C will remain unchanged.

(a)	Initial Population Abundance	Final Abundance	% Population Change [(Final–Initial) / Initial] × 100
Strategy A			
Strategy B			
Strategy C			

(b) Probability of Decline	Strategy A	Strategy B	Strategy C
>100			
> 50			
0			

For Tables 14.8 to 14.9 change several of the parameters together. Record each of the parameters you use in each simulation.

Table 14.8. First set of changes to the model.

Growth rate _____
Dispersal distance _____
Carrying capacity of Philip Flat population _____

(a)	Initial Population Abundance	Final Abundance	% Population Change [(Final–Initial) / Initial] × 100
Strategy A			
Strategy B			
Strategy C			

(b) Probability of Decline	Strategy A	Strategy B	Strategy C
>100			
> 50			
0			

Change several of the parameters together. Record each of the parameters you use in each simulation.

Table 14.9. Second set of changes to the model.

Growth rate _____
Dispersal distance _____
Carrying capacity of Philip Flat population _____

(a)	Initial Population Abundance	Final Abundance	% Population Change [(Final−Initial) / Initial] × 100
Strategy A			
Strategy B			
Strategy C			

(b) Probability of Decline	Strategy A	Strategy B	Strategy C
> 100			
> 50			
0			

Glossary

Age-structured model A mathematical model of a population, in which demographic properties of individuals (such as survival and fecundity) change with age.

Allee effects Ecological effects that lower the growth rate of small populations. For example, individuals may be less likely to find a mate in a small population or may be more vulnerable to predators.

Carrying capacity The maximum number of individuals a habitat can support.

Contest competition Competition that results in unequal sharing of resources. In this type of competition, there are always some individuals that get enough to survive and reproduce by successfully excluding other individuals from resources.

Culling Removing individuals from a population.

Demographic stochasticity/variability Random fluctuation in the number of survivors or offspring because the number of individuals in an age class is a whole number. For example, if the survival rate for an age class is 0.5, on average, half the individuals in that age class will die. However, if there is a finite number of individuals in an age class, the actual number that die may not be exactly the same as the average. This becomes important when there are few individuals, as there might be for an endangered species. Compare with **Environmental stochasticity/variability**.

Density dependent When the value of some variable is affected by the density of something. Often describes the situation in which population growth rate decreases as density of a population increases.

Edge effects The differences in habitat characteristics of an area near the border of a new habitat type. Often used to refer to the differences in forest structure and population density in the area of forest adjacent to degraded land.

Emigration The movement of individuals out of an area or population.

Endemic Native to a particular region.

Environmental stochasticity/variability Variability in demographic rates caused by random variation in the environment, such as weather. Compare with **Demographic stochasticity/variability**.

Exponential growth A constant rate of population growth per individual in the population. Thus, the more individuals in the population, the faster the population size grows.

Extinction vortex The tendency of a population to continue declining toward 0 once its numbers fall below a certain value.

Immigration The movement of individuals into an area or population.

Maximum sustainable harvest/yield The number of individuals that can be removed from a population without threatening the survival of the population.

Metapopulation A collection of populations of the same species that interact through exchange of individuals, or dispersal.

Minimum viable population size The smallest number of individuals in a population that will ensure the persistence of the population over a given time period.

Overexploitation Harvesting a resource at levels that are not sustainable, putting it at risk of regional or global extinction.

Population A group of individuals of the same species, living in a given area at a given time.

Population viability analysis (PVA) Demographic modeling used to determine whether a population is likely to persist or become extinct.

Scramble competition Indirect competition between individuals through the sharing of resources. As new competitors are added, all competing individuals become deprived of the same amount of resources. This type of competition is common in species competing for a dispersed resource, where one individual cannot monopolize access to a resource patch.

Sensitivity analysis A test that ranks the importance of habitat characteristics in influencing the growth and stability of a population.

Standard deviation A measure of the dispersion among a set of measurements. Calculated as the square root of the average squared difference between the mean and observed measurements.

Sterilization A method to prevent individuals from reproducing. Can be used to slow a population growth rate without culling animals.

Stochasticity Random variation (determined completely by chance).

Translocation The movement of individuals from one area to another. Translocation generally removes individuals from a large population or overpopulated area to a smaller population.

Underexploitation When a resource is harvested at a rate lower than the maximum sustainable harvest rate.

Using RAMAS EcoLab

Installation

The program requires an IBM-compatible computer running Windows 95, Windows NT 4.0, or later. The program will not work under Windows 3 or 3.1.

Memory: The computer should have at least 16 megabytes of memory. More memory would improve performance.

Processor: The program will run on an 80486 processor, although we recommend a Pentium or faster processor.

Hard disk space: The program requires approximately 2 megabytes of hard disk space.

If you received the program on floppy disk(s), you must install the program on a hard disk. Put the disk (#1, if there are more than 1) in drive A. Select "Run" from the Start menu, type

a:\setup.exe

and press Enter. Follow the instructions on the screen.

If you received the program on a CD-ROM disk, you must install the program on a hard disk. Put the CD-ROM disk in the CD-ROM drive. The installation should start running automatically. If it does not, select *Run* from the *Start* menu. Type

x:\setup.exe

where x is the letter of the CD-ROM drive, and press Enter. Follow the instructions on the screen. Store the distribution disk(s) in a safe place in case any of the program files are accidentally deleted.

Note: Read the file README.TXT for last-minute hints and corrections.

Uninstalling RAMAS EcoLab

You might want to uninstall RAMAS EcoLab when you change computers or upgrade to a newer version of the program. You can do this by selecting *Uninstall* from the RAMAS EcoLab group under *Programs* in the *Start* menu. Note that this will delete all files that came with the program (including sample files). If you have made changes to any sample files that you'd like to keep, first copy them to a folder other than the folder where you initially installed RAMAS EcoLab (usually C:*Program Files**EcoLab*).

Running the Programs

RAMAS EcoLab will be installed under your computer's *Program Files* folder. Double-click on the RAMAS EcoLab icon on your desktop to start the program. Press the F1 button on your keyboard for help. You can also start RAMAS EcoLab from the RAMAS EcoLab group under *Programs* in the *Start* menu, or by double-clicking on the icons of associated data files (.SP, .ST and .MP).

Double-click on the RAMAS EcoLab icon on your desktop to start a shell program that provides access to all programs of RAMAS EcoLab (as below).

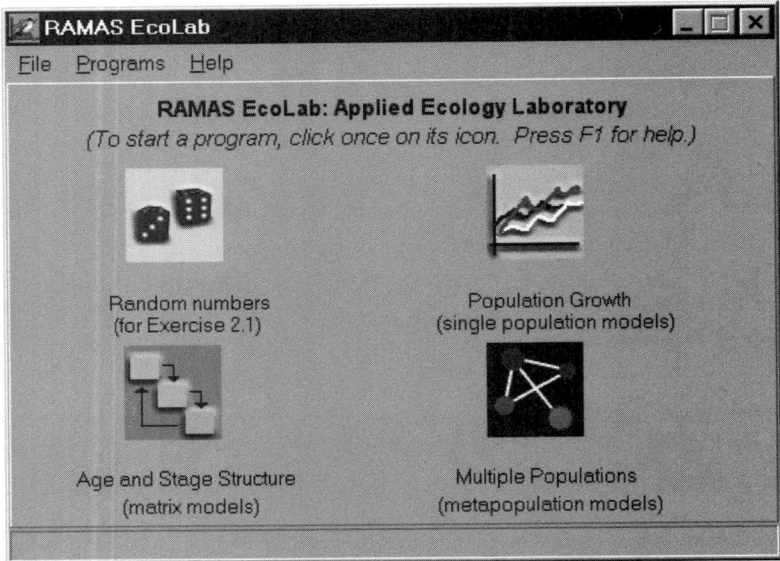

Opening this shell allows you to start any of 4 programs. You may start a program by clicking on its icon or selecting the name of the program from the *Programs* menu. Below is a description of each of the programs next to its icon.

 Random Numbers lets you sample uniform random numbers. It gives a pair of uniform random numbers every time you click a button.

The remaining three programs let you build models:

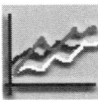 *Population Growth* lets you build single population models with no age or stage structure (i.e., unstructured, or scalar models). These models can have variability or density dependence.

 Age and Stage Structure lets you build single population models with age or stage structure, such as Leslie matrix models and stage matrix models. These models can have variability and density dependence, as well as harvesting.

 Multiple Populations lets you build metapopulation models with spatial structure. These models can have variability, density dependence, and migration among populations.

The use of these three programs is very similar. Each program's main window consists of (1) title bar, (2) menu bar, (3) tool bar, (4) model summary, and (5) status bar.

(1) *Title bar*: At the top of the window is the title bar with the program name. On the title bar, at the upper-right corner of the window, are three buttons for minimizing, maximizing □ (or restoring to original size), and closing the main program window. Clicking the close button will terminate the program.

(2) *Menu bar*: Below the title bar is the menu bar, which includes six menus:

Click on one of these 6 words to open the pull-down menu. Alternatively, you can press the *Alt* key in combination with the underlined letter in the menu name. For example, pressing *Alt*-M will open the Model menu.

File menu is used to open or save model files. *View* menu is used to set display options. Selecting each item in the Model menu opens a dialog box that includes a group of model parameters. *Simulation* menu is used to run a simulation. After running a simulation, selecting each item in the *Results* menu displays one type of model result. The entries listed under *Model* and *Results* menus depend on the program. In each program, click *Help* to learn more about the operation of the program or click on the question mark icon found in some of the windows.

(3) Toolbar: Below the menu bar is the toolbar, which includes four buttons that can be used as shortcuts to access the following functions found under the File menu:

New (start a new model; same as pressing *Ctrl-N*)
Open (open an existing model; same as pressing *Ctrl-O*)
Save (save the model in a file; same as pressing *Ctrl-S*)
Exit (close the program; same as pressing *Alt-X*)

(4) Model summary: The largest part of the main program window contains a summary of the model. Depending on the program, this summary can take two forms:

(a) Text , including title and comments (from the *General Information* dialog), the number of replications, time steps, stages, and populations or,
(b) Map of the metapopulation

(5) Status bar: At the bottom of the main program window is the status bar, which displays information about what the program is doing, as well as hints.

You can resize the program window by clicking on the lower-right corner of the window and dragging. Some of the selections in the menus of a program (for example, *Run*) are procedures, and selecting them will make the program start computing. Others are dialog boxes for entering input parameters or displaying results. When you select one of the dialog boxes for input, the program will display a template on which you can type the values of the various parameters. After you enter your parameters, click OK. If you want to leave a dialog box without making any changes to the input data, click Cancel. The changes you have made since you opened the dialog box will be ignored. For help about input parameters, click Help (or press F1).

Below, we discuss the general features of these programs.

Loading input files

In each program, you can load sample files. To do this, select Open from the File menu (or, press Ctrl-O), type in the filename, or select a file by clicking.

Saving models and results

In each program, you can save a model you have created or modified. To do this, select Save as (to save a model with a different name) or Save (to save with the same filename) from the File menu. If you have already run the model, the results will also be saved.

Entering data

Within input windows under the Model menu (such as General Information), you can type in parameter values, as well as a title and comments. In all subprograms, the number of time steps (duration) and the number of replications are entered in General Information.

Setting replications to 0 is a convenient way of making the program run a deterministic simulation, even if the standard deviation of the growth rate is greater than 0. When the number of replications is specified as 0, the program assumes a deterministic simulation and ignores parameters related to stochasticity. These parameters include the standard deviation matrix for age- or stage-structured models, and the parameters that are dimmed (not available for editing) in other input windows.

After editing an input window, click OK (Note: Don't press Esc or click Cancel to close an input window; use these to undo the last change you have made in this window.)

Erasing all input data and all results

To erase all input parameters and all results of a model, simply start a new model. You can do this by selecting New from the File menu.

Using the help

The function key F1 provides access to a context-sensitive help facility. You can press or click the Help button anytime to get help about a particular window. In the help facility, click on a topic and click Open.

Running a simulation

After you have loaded a file, or created a model, you can run a simulation by selecting Run from the Simulation menu (or by pressing Ctrl-R). When the simulation starts, the program will open a Simulation window.

There are several controls on the toolbar at the top of the Simulation window. The first two buttons on the left (right under the word *Simulation* in the title) allow you to choose the simulation display (what to display during a simulation). By the default, the program will display trajectories 📄 or the metapopulation map ●, depending on the program.

For unstructured and age- or stage-structured models, the program will display the population trajectory simulated by each replication. For metapopulation models, the program will display a map of the metapopulation and will update the map at every time step. The display of trajectories or maps may slow down the program. To turn off the display, click the first button from the left 📄 on the toolbar. This will display only text (title, comments, and other parameters) during a simulation. This allows the simulation to be completed faster. For more information, click the help button ❓ on the toolbar of the Simulation window.

When a simulation is completed, you will see *End of simulation* at the bottom of the window. Close the Simulation window (click on the X in the upper-right corner) to return to the main window. Once you return to the main window, you cannot go back to the display of individual trajectories (unless you run the simulation again).

Viewing and printing results

To view or print the results of a simulation, select one of the entries under the Results menu. This will open a window and display a graph. On top of the window is a series of buttons that will

    Show a plot (display the result graphically, which is the default)

    Show numbers (display the result as a numerical table)

    Open a window for changing the scale and titles of the graph

    Save the result as a disk file

    Print the result (plot or text) on the default Windows printer

    Copy the result to the clipboard, for pasting into another application

    Display help for the particular result

For more information, click the Help button (or press *F1*) and then click on *Copying, saving and printing results*.

When a graph is displayed, the axes may have the letters *k*, *m*, or *b*. These indicate the multiplication factors:

k	× 1000
m	× 1000,000
b	× 1000,000,000

Thus 2.50*k* means 2500 and 0.2*m* means 200,000.

Altering scales, labels, and titles in graphs

RAMAS EcoLab tries to pick sensible scales for graphs, but you may want to alter them. You may also want to change the title and the axis labels. You can do this by clicking the *Scale* button in the window that contains your graph (the third button from the left). To change the title, the *x*-axis label, or the *y*-axis label, simply click the mouse in the box with the element to be changed, and use the keyboard to edit it. If you want to change the scale of either the *x*-axis or the *y*-axis, you must make sure that *Autoscale* is not selected (i.e., the checkmark must not be there). If it is selected, simply click on *Autoscale* and the checkmark will disappear. At this point, you may click on the number to be changed and edit it.

Some general graphing guidelines are
- If you are comparing the results from several different simulations with slightly different parameters, it's best to plot them all on graphs with the same scale if possible.
- Don't use too many decimal places for axis numbers; it doesn't make the graph any more informative (1.3 is better than 1.30000 on an axis).
- Axis labels are important. Without labels, no one looking at the graph will understand what it is supposed to show. A label like *Abundance of spotted owls* is more useful than *Abundance*.
- Too much information can be as bad as too little. Decide which of your results are really important, and present these as graphs. Unimportant results will only confuse your audience.

Exiting the program

To exit from one of the subprograms, select *Exit* from the *File* menu (Important: Remember to save your results before you exit the program).

Technical support

User support from Applied Biomathematics is limited to technical aspects of using the program. The RAMAS home page has a list of frequently asked questions. If you want to contact us, please indicate the program and model you are using, describe the question or difficulty in detail, and if possible, attach a copy of the input file you were working on.

      homepage: http://www.ramas.com
      e-mail: EcoLab@ramas.com
      address: 100 North Country Road, Setauket, N.Y. 11733 USA